Ophthalmic Lenses: Dioptric Formulae for Combined Cylindrical Lenses : The Prism-Diotry, and Other Optical Papers - Primary Source Edition

Charles Frederick Prentice

DR. SWAN M. BURNETT'S MODELS, DEMONSTRATIVE OF CYLINDRICAL REFRACTION.

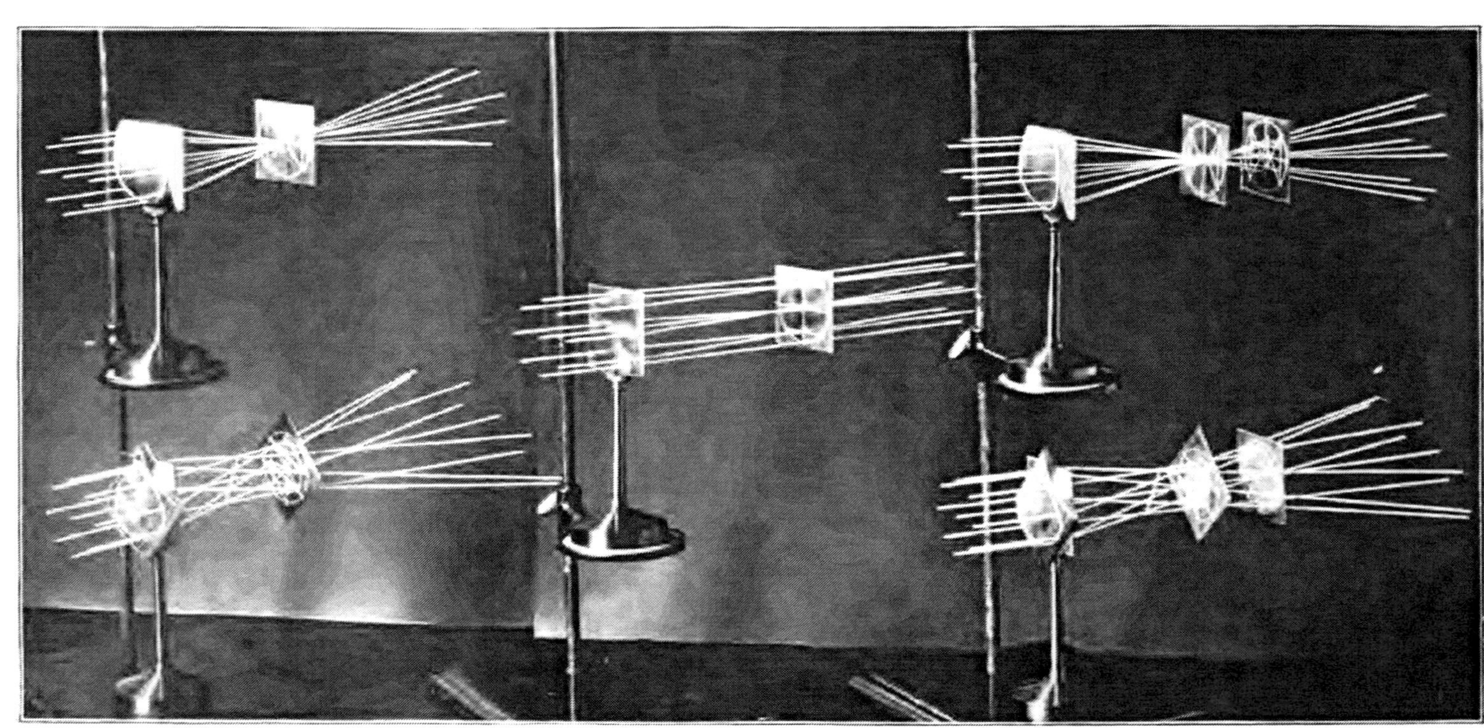

Equal Cylinders, 10 Dioptries, axial deviation 90°.

Equal Cylinders, 10 Dioptries, axial deviation 45°.

Plano-convex Cylinder, 7 Dioptries, axis vertical.

Unequal Cylinders, 7 and 10 Dioptries, axial deviation 90°.

Unequal Cylinders, 7 and 10 Dioptries, axial deviation 45°.

OPHTHALMIC LENSES

DIOPTRIC FORMULÆ FOR COMBINED CYLINDRICAL LENSES

THE PRISM-DIOPTRY

AND OTHER

OPTICAL PAPERS

WITH ONE HUNDRED AND TEN ILLUSTRATIONS

CHARLES F. PRENTICE, M.E.
OPTICIAN

PUBLISHED BY
THE KEYSTONE
THE ORGAN OF THE JEWELRY AND OPTICAL TRADES
19TH & BROWN STS., PHILADELPHIA, U.S.A.

OPHTHALMIC LENSES

DIOPTRIC FORMULÆ FOR COMBINED CYLINDRICAL LENSES

THE PRISM-DIOPTRY

AND OTHER

OPTICAL PAPERS

WITH ONE HUNDRED AND TEN ORIGINAL DIAGRAMS

BY

CHARLES F. PRENTICE, M.E.
"OPTICIST"

PUBLISHED BY
THE KEYSTONE
THE ORGAN OF THE JEWELRY AND OPTICAL TRADES
19TH & BROWN STS., PHILADELPHIA, U. S. A.
1900

PREFACE

In the Treatise on Ophthalmic Lenses an endeavor is made, by means of graphic and analytic methods, to guide the novice upon a path by which he may easily acquire a knowledge of lenticular refraction without recourse to mathematics. The text is, therefore, confined to the description of a series of diagrams, by aid of which the principles involved are presented in their natural order of succession.

Shortly after publication of the above treatise, Dr. Burnett, of Washington, D. C., kindly suggested the production of plastic models of combined cylindrical lenses, by placing an incomplete set of models of his own conception with the author for further elaboration. With it came the request, if possible, also to produce two combinations in which the cylinders were to be united at angles other than right angles. As a result of the author's research, during the time devoted to the construction of the latter more especially, and with a view to establish confidence in the precision of these models, a complete mathematical demonstration of the refraction by combined cylindrical lenses was first presented in 1888.

The purpose of republishing the Dioptric Formulæ, in Section II, is to present the important results then obtained, with as much abridgment of the calculations as permissible.

While the diagrams have been prepared with great care, yet they are somewhat at variance with the laws of true perspective, owing to the author's desire to strictly preserve therein all important circles and right angles referred to in the text.

Among the many attempts to solve this problem, this is the only one whose formulæ contain exclusively the known quantities, namely, the angle between the cylinders and their foci or powers. Furthermore, this is the only solution which has ever disclosed the sixteen laws inherent in a pair of superposed cylindrical lenses—a fact which may be easily verified by comparing this solution with those of Donders, Reusch, Heath, Jackson, Hay, Weiland, Suter and Thompson, whose formulæ are also far less simple.

The main object in republishing the Prism-Dioptry and other original papers, in Section III, is to present them collectively to the student of optometry, who in most instances will now find it very difficult to gain access to the numbers of the journals in which the various original articles have appeared during the past ten years. This action seems to be further justified by the frequent inquiries made for reprints, which have long since been exhausted, and also for the reason that American manufacturers have, since 1894, universally adopted the prism-dioptral system for their entire marketable product of prisms. This should also prove an incentive for all who make use of such prisms to become thoroughly familiar with the character and capabilities of the prism-dioptry.

The manuscripts on prisms were generally approved, before original publication, by Dr. Burnett, to whom the author is indebted for having called attention to the necessity for a new system of numbering prisms, and, indeed, also for having suggested the name of its unit, the prism-dioptry. Therefore its original form of spelling is herein retained. For the sake of greater clearness it has also been deemed advisable to make several important additions to the original text, which, in its present form, may at least be considered authentic with reference to modern ophthalmic prisms.

The appended revised papers will, it is hoped, prove of permanent interest, as they contain features of scientific value not to be found in any hand-book or treatise on optics.

CHARLES F. PRENTICE

NEW YORK, August 15, 1900.

GENERAL CONTENTS

SECTION I

SECTION II

SECTION III

CONTENTS

OPHTHALMIC LENSES

SECTION I

* See Dioptric Formulæ for Combined Cylindrical Lenses.

CONTENTS

DIOPTRIC FORMULÆ FOR COMBINED CYLINDRICAL LENSES

SECTION II

LIST OF PLATES

FRONTISPIECE

PLATES I AND II

PLATES III AND IV

CONTENTS

THE PRISM-DIOPTRY
AND OTHER OPTICAL PAPERS

SECTION III

SECTION I

OPHTHALMIC LENSES

THEIR REFRACTION AND DIOPTRAL FORMULÆ

WITH THIRTY-SIX ORIGINAL DIAGRAMS

REFRACTION.

§ 1. The change wrought in the direction of oblique rays of light, on their passage from one transparent medium to another of different density, is called Refraction. As our proposed treatment of its manifestation by lenses is to be strictly elementary, we must first define the law of refraction, at least so far as applied to parallel rays, in air, impinging upon and passing through transparent optical glass. In the accompanying diagram, Fig. 1, a piece of glass of considerable thickness, having parallel surfaces *a–b* and *c–d*, is presented as an isolated vertical section *a–b–c–d*, which is exposed to the oblique ray *i* in the same plane of the surrounding air.

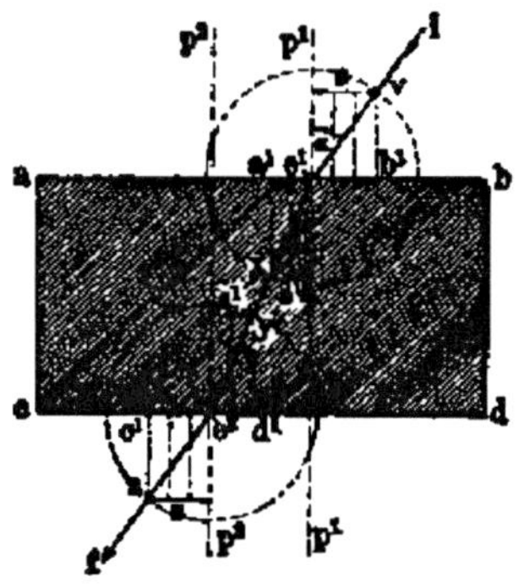

Fig. 1.

For convenience we shall term the ray prior to its contact with the glass, the incident ray, *i*; the ray during transit within the glass, the refracted ray, e^1 e^2;* and the refracted ray after exit, the final ray, *f*.

* The use of superior indices will not prove conflicting, as algebraic values are excluded.

§ 2. Refraction manifests itself by an acute bend in the direction of an oblique ray of light, *i*, at the point of incidence, e^1, in passing from one conducting medium to another, *a–b–c–d*, of different density. Hence, a ray passing from one into and through another medium is bent both at the point of entrance e^1 and of exit e^2.

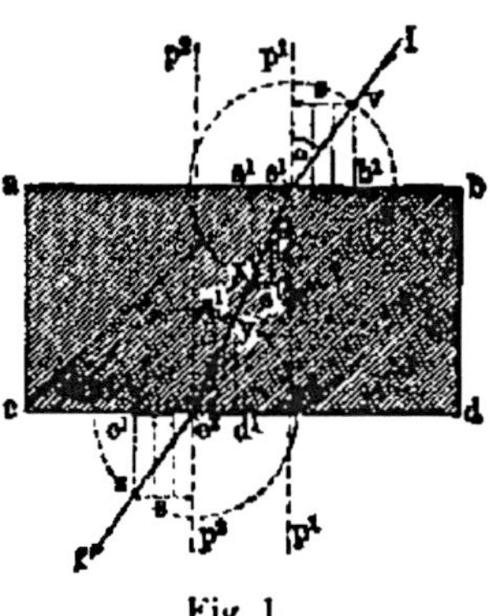

Fig. 1.

By virtue of the deflection or bend alluded to, the incident ray, *i*, must include a different angle, α, with the perpendicular p^1 from that, β, of the refracted ray e^1 e^2; and it is by the trigonometrical values s and s^1 of these angles, which have been found to bear a constant proportion to each other, that we are enabled to give expression to the amount of deflection sustained by a ray in passing from one medium to another.

§ 3. Experiment has shown that the proportion $\frac{s}{s^1}$ remains a constant value for any obliquity of a ray incident to the same medium, and yet, that it possesses a different value by substituting one medium for another.

It has therefore been considered expedient to establish the value of $\frac{s}{s^1}$ for all transparent media, in the specific case of a ray passing from *air* into them; such values being known as the refractive indices of the substances.

To illustrate the graphical method by which we may arrive at the direction of the refracted ray, when the index of refraction and the direction of the incident ray are known, we shall select the index for crown glass = 1.5, by introducing the proportion $\frac{s}{s^1} = \frac{3}{2} = 1.5$ in the construction as follows:

After erecting the perpendicular p^1, take from a scale of equal parts the value for $s = 3$, and transfer it between e^1 and b^1 beneath the ray i, upon the line e^1 b.

In the same manner transfer the value for $s^1 = 2$ from e^1 to a^1, upon the line e^1 a, and in both of these points, b^1 and a^1, erect perpendiculars. The perpendicular at b^1 will intersect the ray, i, at a point, v, which limits the radius of a circle drawn from e^1 as a center; and by the circle's intersection with the perpendicular at a^1 the point, x, defining the direction of the ray, e^1 e^2, is fixed.

§ 4. As a ray of light is propagated backwards or forwards on the same path, the index of refraction from a denser medium into air is the inverse proportion from that of air into the medium, hence $\frac{s^1}{s}$ is the proportion by which the direction of the final ray, f, is to be determined when the direction of the ray, e^1 e^2, is known.

We therefore erect at e^2 the perpendicular p^2 and transfer the value of $s^1 = 2$ beneath the ray e^1 e^2 from e^2 upon the line e^2 d; likewise the value for $s = 3$ from e^2 upon the line e^2 c, and erect, as before, in the points d^1 and c^1 the perpendiculars.

The perpendicular at d^1 will intersect the ray, e^1 e^2, at a point, y, limiting the radius of a circle from the point e^2; the point, z, at the circle's intersection with the perpendicular in c^1 establishing the direction of the final ray, f.

As p^1 and p^2 are parallel, from the construction it follows that the ray f is parallel to i, and therefore of the same direction.

PRISMS.

§ 5. Pursuant to the spirit of our intention to avoid mathematical formulæ, we shall seek to arrive at a conclusion respecting the deflection incurred by a ray in passing through a medium with oblique plane surfaces, confining ourselves as before to isolated vertical sections.

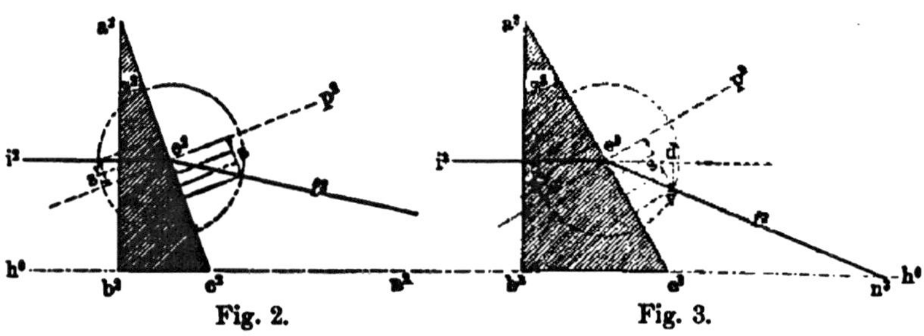

Fig. 2. Fig. 3.

Specifically, we shall select two right-angled prisms of varying angles, a^2 and a^3, with the rays i^2 and i^3 incident perpendicularly to the vertical sides a^2 b^2 and a^3 b^3, so as to avoid refraction on the incident sides, as shown in the vertical sections, Fig. 2 and Fig. 3, respectively. At e^2 and e^3 the rays i^2 and i^3 suffer refraction in the proportion $\frac{s^1}{s} = \frac{2}{3}$, according to § 4, and which, if carried out in the construction, as before indicated, determines the directions of f^2 and f^3, respectively, as shown.

In the future we shall have occasion to refer to the line dv, which is the perpendicular from v upon a line coincident with the ray i^3 when the latter is parallel to the base b^3 c^3 of the prism, Fig. 3. Under such circumstances the displacement dv of the final ray f^3 is associated with a mathematical dependency upon the angle, a^3, of the prism, and the index of refraction $\frac{s}{s^1}$. See page 108.

From the construction it follows that the final ray f^3 (Fig. 3) intersects the horizontal line h^0 at n^3; and f^2 (Fig. 2) at a more distant point, n^2, not shown. By a comparison of the prismatic section Fig. 2 with Fig. 3, we observe that by a decrease of the angle from a^3 to a^2 the perpendicular p^2 has a greater tendency to parallelism with the horizontal line h^0 than p^3. Such parallelism being realized—when $a^2\ c^2$ is parallel to $a^2\ b^2$ or $a^2 = 0°$ —would result in the value s^1 vanishing in the incident ray i^2, and s in the final ray f^2, by virtue of the decrease of the angles of incidence and refraction in the proportion 2 to 3, thus establishing the coincidence of the incident and final rays, and placing the point (n^2) of intersection at infinity respecting the horizontal line h^0.

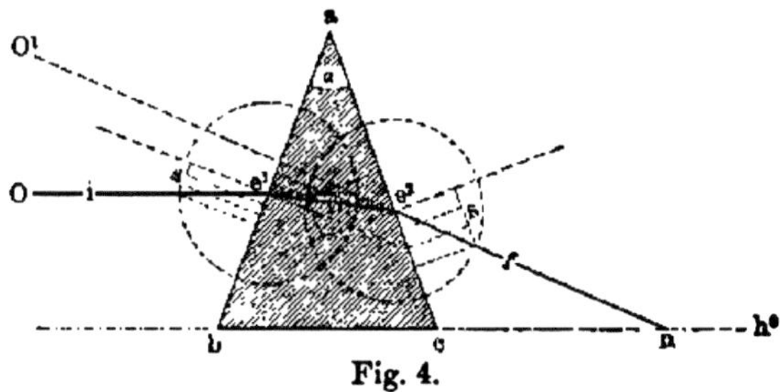

Fig. 4.

§ 6. In general we may therefore be permitted to assume that the greater the angle, a, of obliquity of the surfaces (Fig. 4) the greater will be the deflection of the final ray f, and the closer to the base bc of the prism will be its intersection n with the horizontal line h^0. In looking through a prism the refraction manifests itself by an apparent change from the true position of an object O, to that of its image O^1, when viewed from the point n. The author's suggestion that this phenomenon should form the basis of comparison in measuring prisms was adopted by American manufacturers in 1895. The unit of prismatic refraction is equal to a deflection (dv, Fig. 3) of one centimeter at one meter's distance, and is called the prism-dioptry.

§ 7. By confining our observations to the relative directions of the incident and final rays, we may easily memorize the law of refraction, for a medium included within plane surfaces, in the following manner:

1, a. The direction of a ray remains unchanged in passing through opposite parallel surfaces of a transparent medium, or

b. The incident ray *i* and the final ray *f* are parallel when the former is projected obliquely upon a transparent medium included within parallel surfaces.

2, a. The direction of a ray is changed in passing through opposite oblique surfaces, by a deflection of the final ray *f* toward the region of their greatest distance apart, or

b. The incident ray *i* and the final ray *f* are oblique when the former impinges upon a transparent medium included within oblique surfaces, or

c. The apex of the angle formed by an obliquity of the incident and final rays is always directed toward the apex of the angle of obliquity of the surfaces.

The law of refraction (2) finds its graphical demonstration in the following figures wherein we have introduced the medium glass as being intersected by imaginary vertical and horizontal planes, *V* and *H*, coördinate at the point of exit e^2 for the final ray *f*.

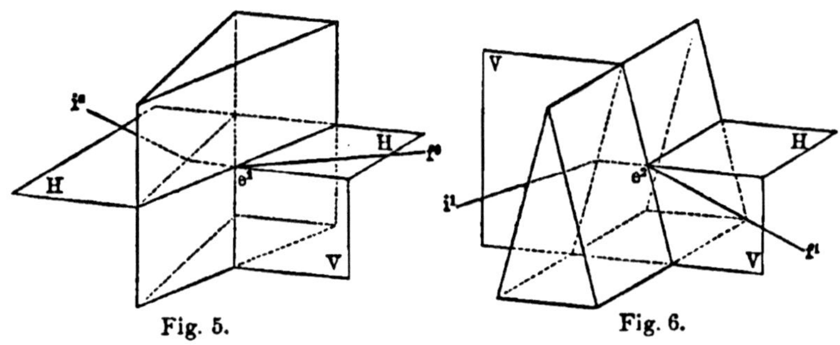

Fig. 5.

Prism, Base vertical; Refraction horizontal.

Fig. 6.

Prism, Base horizontal; Refraction vertical.

§ 8. The Figures 5 and 6 are of particular interest to us, as they illustrate a very vital element in our future consideration of the refraction by cylindrical lenses, namely, that the refraction is strictly confined to the plane whose intersection with the medium defines the obliquity of its surfaces. Thus, for an obliquity of the surfaces in the horizontal plane *H* (Fig. 5), we find the refraction active in the horizontal plane (i^0 to f^0), and for an obliquity of the surfaces in the vertical plane V (Fig. 6), the refraction is active in the vertical plane (i^1 to f^1).

Here, in the sense that the final rays are confined to the plane of incidence, we may term the refraction passive in respect to its right-angled coördinate plane. Thus in Fig. 5 the refraction is passive with regard to the vertical plane, and in Fig. 6 with regard to the horizontal plane.

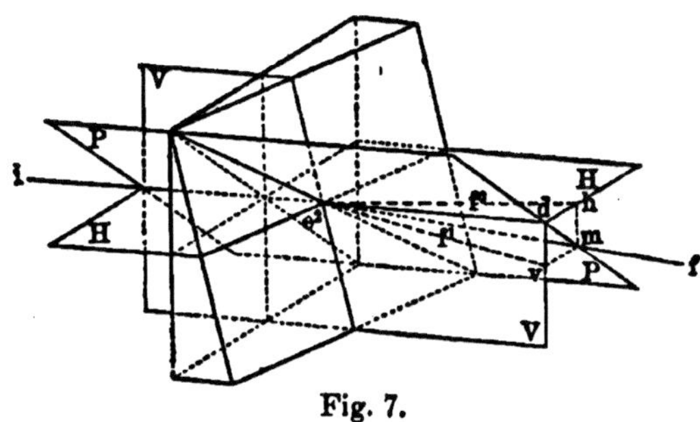

Fig. 7.

Prism, Base oblique; Refraction diametrically opposed.

§ 9. It is evident that the refraction is active in one and passive in the other plane for a medium whose surfaces are oblique in but *one* plane, so that to obtain the refraction active in both fixed planes an obliquity of the surfaces relative to each plane would be necessary. In such a medium (Fig. 7), if we consider the refraction merely with regard to the horizontal obliquity of the surfaces, the final ray would take the direction f^0–h, and, if independently for the vertical obliquity, the final ray would assume the direction f^1–v. Therefore, with due consideration to the obliquity in both planes, the refraction must include both properties of deflection and result in a final ray, f, which is directed to a point, m, defined by projection of the apportioned horizontal and vertical displacements, dh and dv. As this is a prism whose base is really set diagonally to the fixed right-angled coördinate system, the ray f must naturally be refracted in the direction of the greatest distance apart of the surfaces, through the point m, within the diagonally bisecting or oblique plane P.

SIMPLE LENSES.

§ 10. Directing our attention to the effect produced by substituting a segment of a circle for the line $a^2 c^2$ of the original prismatic section (Fig. 2), each succeeding point e^2, e^3, e^4 (Fig. 8) may be considered as one of a prism varying in its angle a^2, a^3, a^4 with that of its predecessor; and if the construction be carried out for each incident ray i^2, i^3, i^4 the corresponding radial lines at the points, e^2, e^3, e^4 in this case substituting the perpendicular p^2 heretofore mentioned, each final ray f^2, f^3, f^4 will be found to intersect an arbitrarily selected base line, h^0, at the respective points n^2, n^3, n^4, to infinity.

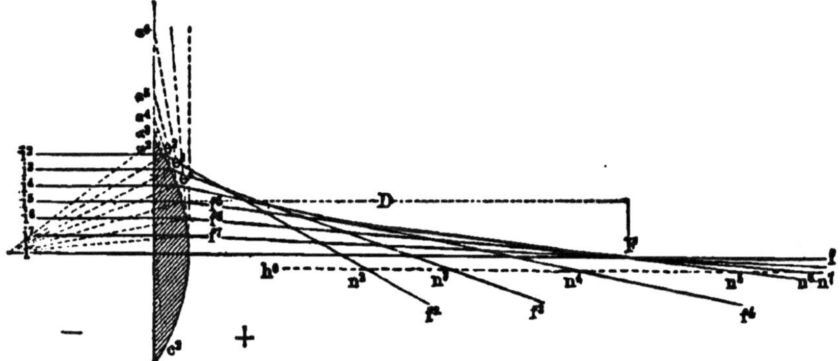

Fig. 8.
Plano-convex section.

In the so-called plano-convex lens, Fig. 8, the converging final rays f^5, f^6, f^7, corresponding to the more central incident parallel rays i^5, i^6, i^7,* establish points n^5, n^6, n^7 to infinity, and possess the remarkable feature of intersecting each other at a common point F, termed the focal point, which is situated upon the central and direct ray i–f. According to § 4, rays emanating from the focal point F, will be emitted as parallel rays i^5, i^6, i^7 and i.

* All future deductions refer exclusively to such rays.

The points n^4, n^3, n^2, toward the lens, correspond to the more eccentric incident rays, and, in the sense that these fail to assist in the harmony of a union of the final rays at the focal point, are to be considered a disturbing element, giving issue to what is termed aberration. In the plano-concave lens, Fig. 9, the final rays f^5, f^6, f^7 are emitted as diverging rays, which may be considered as emanating from the so-called *virtual* focal point F, situated on that side of the section in which the rays are incident.

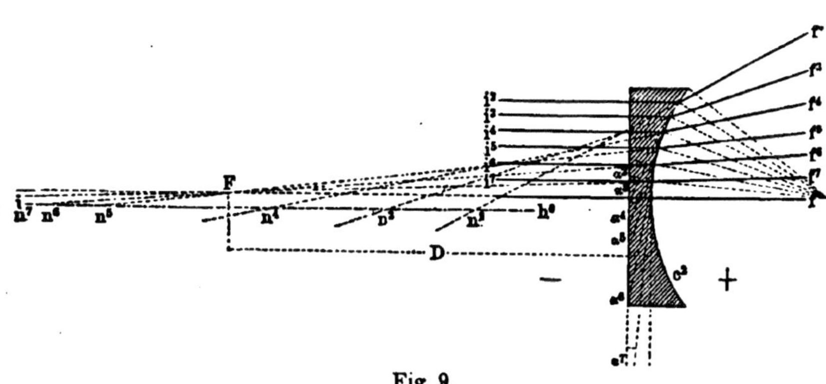

Fig 9.

Plano-concave section.

§ 11. For either of the above lenses it is also obvious that the more acute the curvature of the circle, the greater proportionately will be the angles a^2, a^3, a^4, limiting the obliquity of the surfaces, and the closer to the lens will be the focal point F. Further, as the curvature of the circle is dependent upon the dimensions of the radius, the latter must prescribe the distance, D, of the focal point from the medium or lens whose index of refraction is known. This relationship involves mathematical formulæ for which we refer the reader to special treatises* on the subject

The greater the deflection of the final rays f^5, f^6, f^7, the shorter will be the distance D, or, for an increase in the refraction we have a corresponding decrease of the focal distance. Hence we say that the refractive power of a lens is in inverse proportion to its focal distance.

* Elementary Geometrical Optics, W. Steadman Aldis, M.A., Cambridge, 1886. Hand Book of Optics for Students of Ophthalmology, W. N. Suter, B.A., M.D., New York, 1899.

§ 12. If we express the unit of refraction by the numeral 1, for a lens whose focal distance D is equal to one meter or 100 centimeters, lenses of two, three, or four times the refraction would find the expression of their focal distances in $\frac{1}{2}$, $\frac{1}{3}$, $\frac{1}{4}$, the focal distance of the unit, or 50, $33\frac{1}{3}$ and 25 centimeters, respectively.

The unit above mentioned has been termed the Dioptry, and is now the standard of refraction in optometrical practice. Values beneath the unit are designated as 0.25D.,* 0.50D., and 0.75D., their respective focal distances being four meters or 400 centimeters, two meters or 200 centimeters, and one and one-third meters or $133\frac{1}{3}$ centimeters. Those values which are higher than the unit are expressed in whole numbers, including their intervals as above. See page 47.

§ 13. Assuming the medium to divide the aerial space into negative and positive regions (Figs. 8, 9) as indicated by the sign — (minus) on the incident side of the medium, and the sign + (plus) behind the medium, we shall find the focal point on the positive side for all convex, and on the negative side for all concave lenses.

In this sense the refraction for convex lenses is considered positive, and for concave lenses negative. Hence Fig. 8 is + 1D., and Fig. 9 is — 1D.

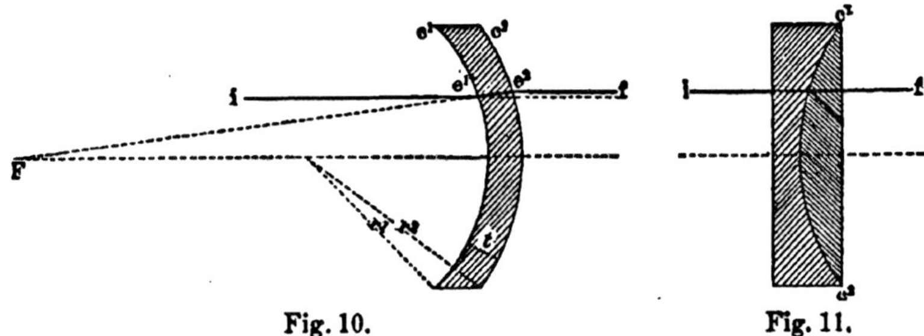

Fig. 10. Fig. 11.

§ 14. By substituting, in Fig. 8, for the plane side, a curvature c^1 concentric with c^2, the refractive effects of the sections, Fig. 8 and Fig. 9, are virtually united, as shown in Fig. 10. Owing to the concave curvature c^1, the incident ray i will assume the direction e^1–e^2, being coincident with the focal point F, which may also be practically accepted as the focal point for

* D here being the abbreviation for Dioptry.

the convex curvature c^2, provided the thickness, t, of the medium is created infinitely small in proportion to the radii r^1 and r^2.

Rays emanating from the focal point F, for a convex curvature c^2, being emitted as parallel rays, § 10, it conditionally follows that the ray f will be parallel to the ray i. The neutralization is the more complete when the curvatures c^1 and c^2 are identical, and are brought in contact as shown in Fig 11, which, however, is a special case.

§ 15. Hence, in a pair of united convex and concave sections of identical curvature, it follows that the effect of the one is neutralized by the other, respecting the existence of a focal point on either side of the medium. This, however, is strictly only true for lenses weaker than 9D.

§ 16. If the opposite curvatures be *unequal*, the final rays will unite at a focal point on that side of the medium which corresponds to the focal point of the more acute curvature.

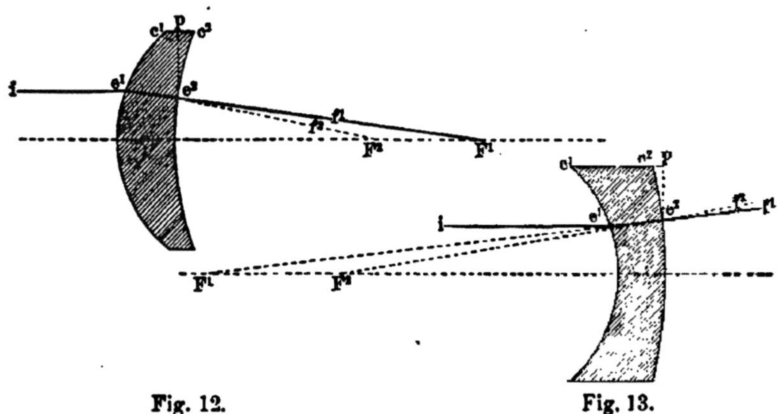

Fig. 12.
Periscopic convex section.
Positive meniscus.

Fig. 13.
Periscopic concave section.
Negative meniscus.

Referring to the periscopic convex and concave sections, Fig. 12 and Fig. 13, respectively, if we consider the refraction merely with respect to the front curvature c^1, disregarding the existence of a terminating back surface, the incident ray i will assume the direction of the ray $e^1 f^1$, toward the focal point F^1, then within the medium.

Accepting the *plane* e^2–p to be the limit of the medium, the ray e^1 e^2

would suffer a second refraction, and result in the ray e^2–f^2, directed to the focal point at F^2.

To eliminate this second or augmented refraction, it would be necessary for the ray e^1 e^2 to impinge upon the back surface c^2 perpendicularly at e^2. A surface effecting this is obtained by giving it a curvature c^2 prescribed from the point F^1 as a center, in which specific event the ray e^1 e^2 traverses the radius of the circle, or the perpendicular at e^2 for the surface c^2, thus fixing the point F^1 as the focal point for the respective periscopic convex and concave sections.

§ 17. Observation of the figures shows that the weaker proportionately reduces the refraction of the more acute curvature, so that the focal point F^1 of the periscopic is at a greater distance from the medium than the focal point F^2 of the plano-convex or concave sections. The more acute the curvature c^2, within the limits of parallelism with the curvature c^1, the more distant will be the focal point F^1 from the medium, so that the total refraction for the respective sections is equivalent to the difference of the apportioned numerals, and bears the sign corresponding to the more acute curvature c^1.

Supposing, in a periscopic convex section, 2.5D. to be the prescribed numeral of refraction for the convex, and 0.50D. for the concave side, the total refraction will be 2.5 — 0.50 = 2D. convex, or + 2D.

Similarly, in a periscopic concave section, 2.5D. concave combined with 0.50D. convex equals 2.5 — 0.50 = 2D. concave, or — 2D.

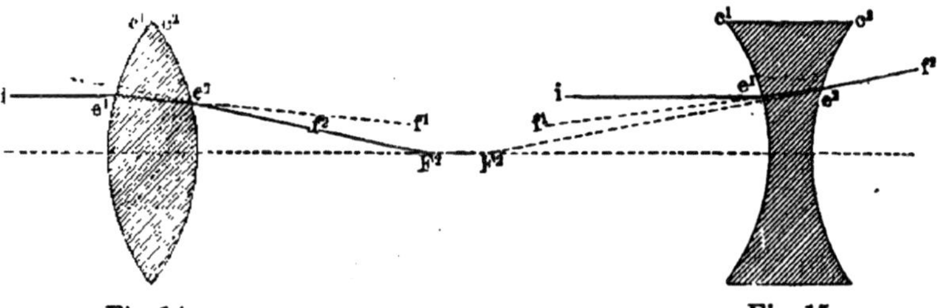

Fig. 14.
Double or Bi-convex section.

Fig. 15.
Double or Bi-concave section.

§ 18. In the bi-convex and bi-concave sections Fig. 14 and Fig. 15 it can be similarly shown that the curvature c^2 increases the refraction of c^1, so that the total refraction is expressed by the sum of the apportioned numerals and bears the sign associated with the respective sections.

Thus in either figure the numeral for c^1 being 1D., and for c^2 being 1.5D., the total refraction will be $1 + 1.5 = 2.5$D.

Convex, or + 2.5D. for Fig. 14, and concave, or — 2.5D. for Fig. 15.

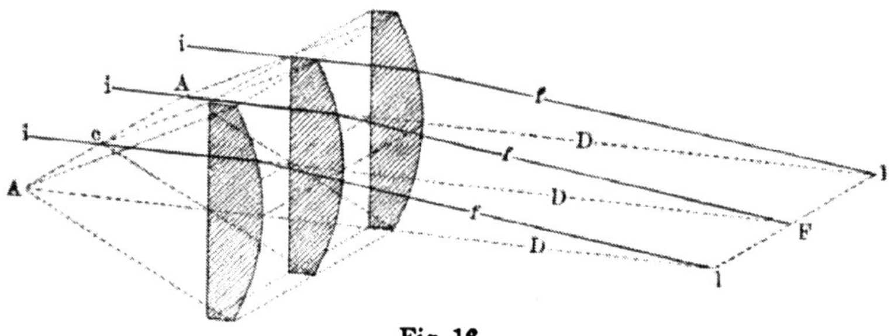

Fig. 16.

§ 19. For a medium (Fig. 16) composed of parallel vertical sections, each adjacent imaginary section has its corresponding focal point at the same distance (*D*) from the medium, so that the refraction for all central incident parallel rays becomes manifest by establishing a succession of these points, resulting in the so-called focal line *l F l*.

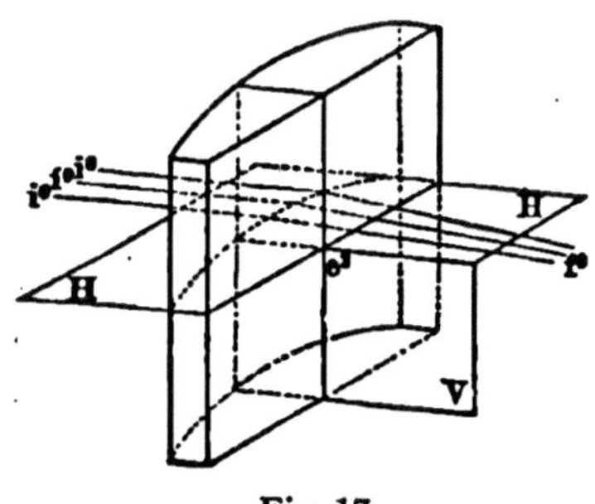

Fig. 17.

Axis vertical; Refraction horizontal.

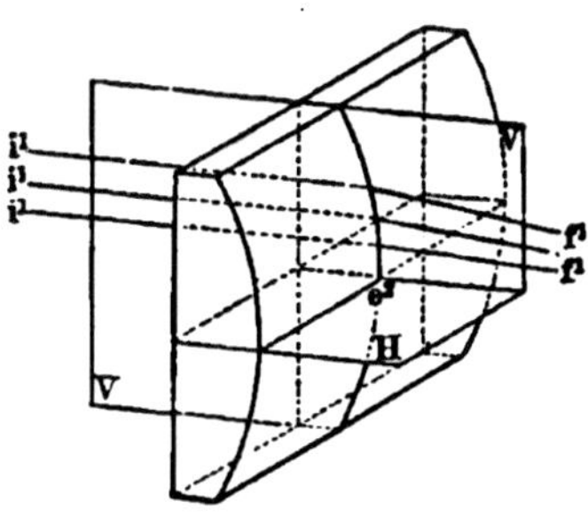

Fig. 18.

Axis horizontal; Refraction vertical

Plano-convex Cylindrical Lenses.

A similar succession of the radial centers (*c*) establishes a line (*A c A*), termed the axis of the so-created cylindrical medium or lens, which is parallel to the focal line *l F l*, and in the same plane.

§ 20. As simple cylindrical lenses have their surfaces of greatest obliquity in the plane which is perpendicular to the axis, we here also find the refraction active in this plane, and passive in the axial or right-angled coördinate plane (see Figures 17-20), wherein, as before, i^0 and f^0 are associated with refraction in the horizontal, and i^1 and f^1 with refraction in the vertical plane.

In a practical experiment in which the lens is held at some distance from the eye, convex cylindrical refraction manifests itself by an apparent *increase*, and concave cylindrical refraction by an apparent *decrease* in the dimensions of an observed object in that plane which is at right angles to the axis. In the axial plane, the refraction being passive, corresponding dimensions remain unchanged.

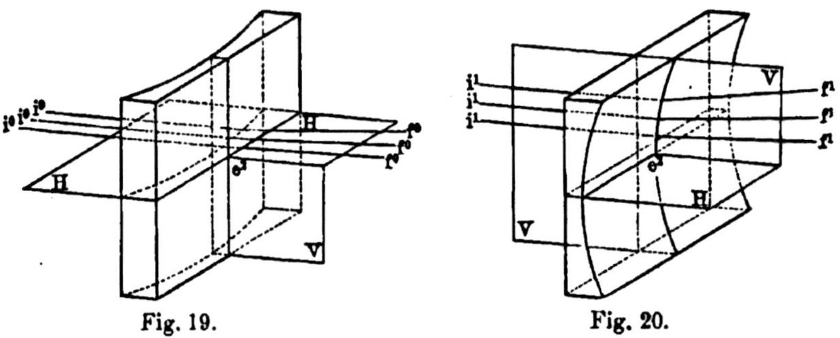

Fig. 19. Fig. 20.

Axis vertical; Refraction horizontal. Axis horizontal; Refraction vertical.
Plano-concave Cylindrical Lenses.

§ 21. To obtain cylindrical refraction of equal amount in both planes, thereby reducing the focal line to a focal point, it would be necessary to combine two identical cylinders, or, to create a single lens whose opposite surfaces are right-angled coördinate cylindrical elements as shown in Fig. 21.

Under such circumstances, however, the focal line $l^1 F^1 l^1$ for the front surface c^1 is slightly closer to the face of the lens than the focal line $l^2 F^2 l^2$ for the back surface c^2. Aside from this, in making a bi-cylindrical lens it is difficult to insure the chief planes of refraction being strictly at right-angles to each other, so that failure in this is certain to increase the aberration.

§ 22. The greater the distance apart of the surfaces, c^1 and c^2, the greater will be the aberrative distance, F^1 to F^2. Yet, as the thickness of the lens may generally be accepted as a vanishing quantity in proportion to the focal distance, we may consider a common focal point to exist for both refracting surfaces.

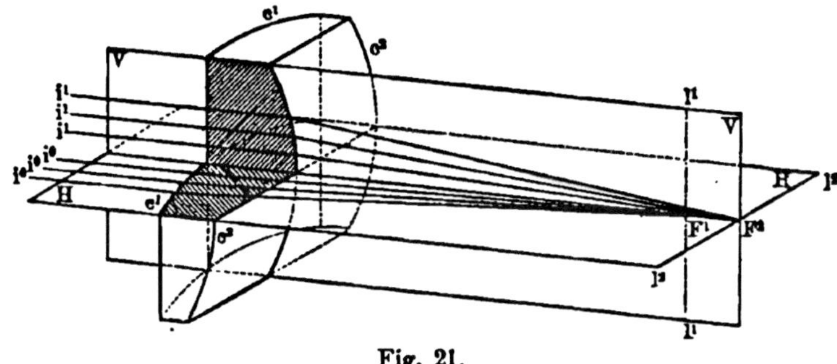

Fig. 21.

Double or Bi-cylindrical Lens.

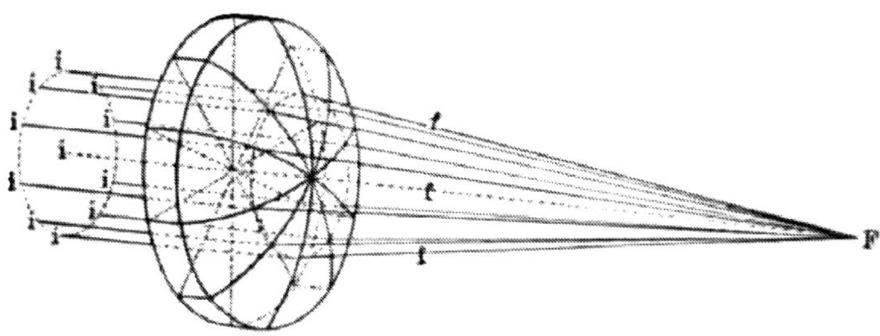

Fig. 22.

Plano-convex Spherical Lens.

§ 23. Practically, however, it would be better to create a single surface capable of producing this amount of refraction in the vertical as well as in the horizontal plane. With this object in view, we shall select the isolated vertical section described in § 10, and cause it to be rotated upon the central incident and direct ray, *i–f*, as its so-called optical axis, whereby a plano-convex spherical lens is obtained. See Fig. 22. Similar rotation of the sections Figs. 9, 12, 13, 14, and 15, would result in the so-created spherical lenses being characterized by the sections employed.

It is evident that the incident and final rays will retain their relative obliquity during the rotation, so that all incident parallel rays have their corresponding final rays in the resulting cone whose apex is at the focal point *F*.

To further illustrate, we may take advantage of § 9 in its application to a medium having only one surface which is spherically curved, and consequently oblique in respect to both right-angled coördinate planes.

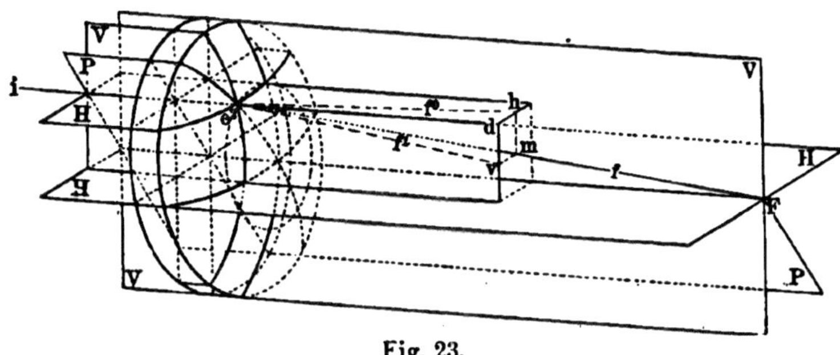

Fig. 23.

In the plano-convex spherical lens, Fig. 23, if we consider the refraction at e^2 of the ray i merely with regard to the horizontal refraction, the final ray would take the direction f^0–h, and, if independently for the vertical refraction, the final ray would assume the direction f^1–v. Therefore, with due consideration to the refraction in both planes, the refracted ray must include both properties of deflection, and result in a final ray f, which is directed to the focal point *F*, through a point m, of the oblique plane *P*, as defined by projection of the apportioned horizontal and vertical displacements dh and dv.

§ 24. Finally, we may therefore conclude that spherical refraction is equivalent to the refraction of right-angled crossed cylinders of identical curvature.

As in spherical lenses the refraction is equably active in any pair of diametrically-opposed meridians, it follows that both the lateral and vertical dimensions of objects seen through them will appear to be enlarged by convex and diminished by concave lenses, when these are held at some distance from the eye.

COMPOUND LENSES.

1. CONGENERIC MERIDIANS (CONVEX).

§ 25. An asymmetrically-refracting or astigmatic lens is one in which the principal diametrically-opposed sections include different degrees of refraction, in contradistinction to those hitherto mentioned, in which uniform refraction took place either exclusively in one meridian, or equally in both meridians.

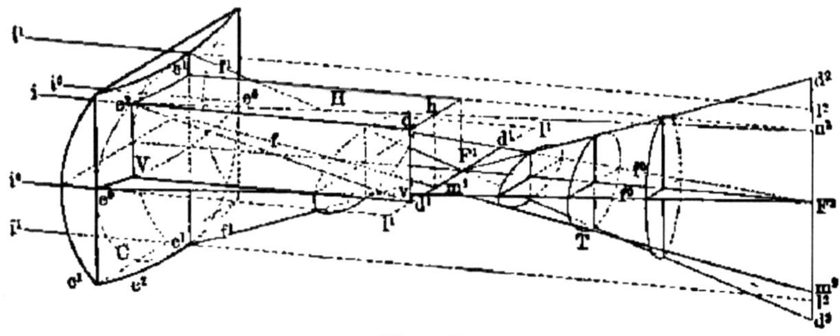

Fig. 24.

Convex Cylindro-cylindrical Lens (+ c^1 axis 180° ⊃ + c^2 axis 90°).

Referring to § 21, Fig. 21, it is evident that the aberrative distance F^1 to F^2 may also be definitely increased by giving different amounts of refraction to the active planes or sections of the combined cylinders. In this event the focal point ascribed to the equally curved bi-cylindrical lens will be effaced, though substituted by a pair of focal lines, whose distance apart will be equal to the difference between the focal distances of the crossed *unequal* cylinders. Thus in the bi-cylindrical lens, Fig. 24, represented as consisting of two crossed convex cylinders (c^1 and c^2) of unequal curvature, l^1 F^1 l^1 and l^2 F^2 l^2 will be the respective *elementary* focal lines. The distance between them (F^1 to F^2) has been termed, by Sturm, the "focal interval."

As the cylinders are of equal length, the focal lines l^1 F^1 l^1 and l^2 F^2 l^2 would also be identical in this regard if the apportioned refractions of the cylinders were considered *independently* of each other.

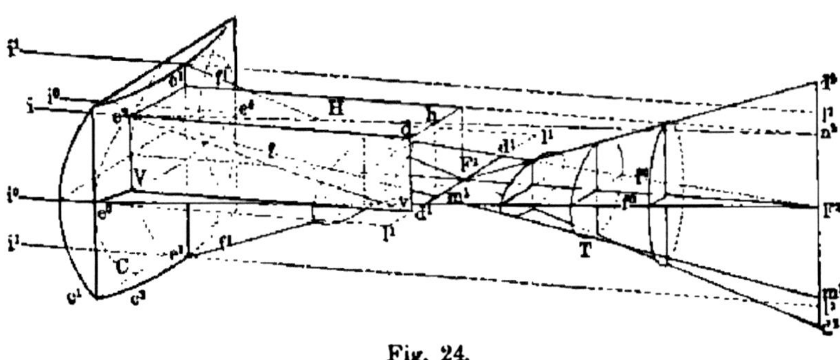

Fig. 24.

Convex Cylindro-cylindrical Lens (+ c^1 axis 180° ⊃ + c^2 axis 90°).

§ 26. The combined refraction of the cylinders, however, definitely modifies this specific condition, and in the following manner:

The outermost incident rays i^{11}, in the central horizontal plane, which would have been directed to the points l^1 and l^1 for the cylinder c^1, will suffer horizontal displacement toward the point F^2, owing to the activity of the refraction in this plane for the cylinder c^2, and so establish points d^1 and d^1 of the focal line l^1 F^1 l^1 for the combined action of the cylinders c^1 and c^2 in the horizontal plane.

Similarly, the outermost incident rays i^1 in the central vertical plane, which would have been directed to the points l^2 and l^2 for the cylinder c^2, will suffer vertical refraction in this plane by the cylinder c^1, which causes the final rays to cross each other at F^1 and to intersect the focal line l^2 F^2 l^2 at the points d^2 and d^2 for the combined action of the cylinders c^1 and c^2 in the vertical plane.

If we consider the refraction at the point e^2 of the circle C for the ray i merely with regard to the horizontal refraction of the surfaces, or the cylinder c^2, the final ray would take the direction e^2–h, intersecting the focal line of the cylinder c^2 at a correlative point n^2; but, as all final rays for the cylinder c^1 above the central horizontal plane intersect the focal line d^1 F^1 d^1, it follows, through introducing the cylinder c^1, that the ray e^2 h must fall subject to the influence of c^1 for the combined action of the cylinders, thus depressing the ray e^2 h from the point h perpendicularly to m^1, and consequently also the point n^2 to m^2 within the focal line d^2 F^2 d^2.

By an analogous reasoning to § 9 we here also find the direction of the final ray f to be determined by projection of the apportioned horizontal and vertical displacements, dh and dv, which are solely dependent upon the active meridians of the cylinders c^1 and c^2.

Increased proximity of the point e^2 to e^0, upon the circle C, will be associated with an increased distance between m^1 and F^1, and with an approach of m^2 toward F^2 for these points of intersection of the final ray f with the respective focal lines F^1 d^1 and F^2 d^2. The reverse is evident for an advancement of e^2 towards e^1. (See Fig. 25a.)

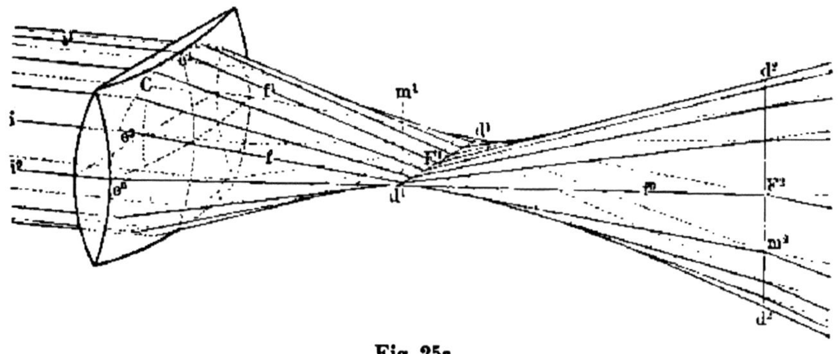

Fig. 25a.

§ 27. The total refraction for all incident parallel rays within the area of the circle C will therefore result in an astigmatic pencil whose focal lines, d^1 F^1 d^1 and d^2 F^2 d^2, are limited as to position and magnitude. This astigmatic pencil, if intercepted at intervals by a transverse perpendicular screen, will project elliptical areas of light whose longest and shortest diameters correspond to the principal meridians of refraction. In the immediate vicinity of F^1, for instance, the ellipses have their longest diameters horizontal; whereas, in the vicinity of F^2, their longest diameters are vertical.

This naturally effects a reversal of the ellipses, respecting their diameters, at some point within the focal interval F^1–F^2; such point being determined where the vertical and horizontal displacements are alike, and the section T, consequently, a "circle of least confusion."

§ 28. Astigmatic refraction in a lens is, however, preferably attained by combining a spherical with a cylindrical surface, the requisite conditions being fulfilled through that increase or decrease of the spherical refraction which is produced by and in the active meridian of the cylinder.

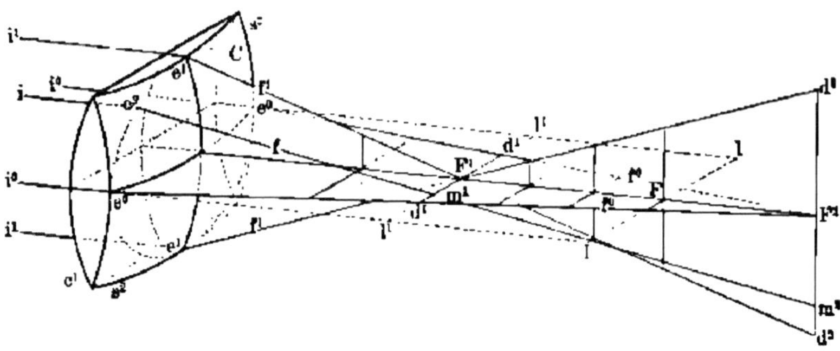

Fig. 25.

Convex Sphero-cylindrical Lens ($+ s^2 \supset + c^1$ axis 180°)—Double Form.

To increase the refraction of a positive or negative spherical lens in one meridian, we may add to it the active meridian of a cylinder bearing the same sign; and to decrease it in the same meridian we may combine it with the active meridian of a cylinder bearing the opposite sign.

(1) The combination of a positive spherical with a positive cylindrical surface would result in the section of *greatest* refraction being *double* convex; and,

(2) The combination of a positive spherical with a weaker or less acutely curved negative cylindrical surface would result in the section of *least* refraction being *periscopic* convex.

Where the aforesaid combinations are spoken of, we shall for convenience, apply to them the terms double and periscopic form, respectively.

§ 29. As the combination of crossed convex cylinders of unequal curvatures gave rise to a pair of focal lines, to the novice it may appear requisite that a focal point and a focal line should exist for the combination of a spherical with a cylindrical surface. We shall consequently endeavor to avert this possible though erroneous impression.

In the convex sphero-cylindrical lens of double form, Fig. 25, if we considered the refraction for each surface independently of the other, we should find a focal point at F^2 for the convex spherical surface, s^2, and a focal line, say at $l\ F\ l$, for the cylindrical surface c^1. Their combination giving rise to augmented refraction in the vertical plane, however, occasions a displacement of the focal line $l\ F\ l$ to the position of $l^1\ F^1\ l^1$.

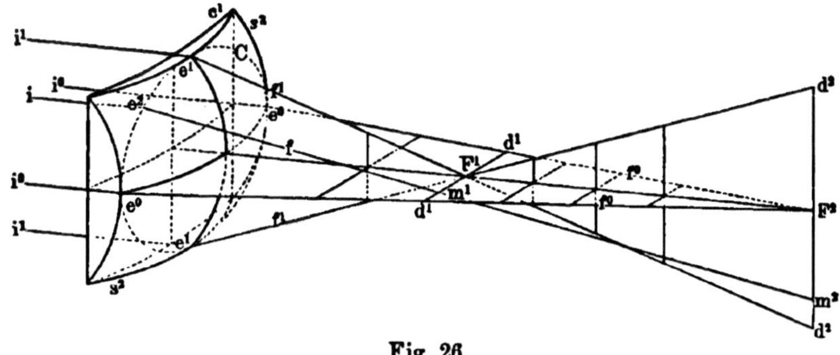

Fig. 26.

Convex Sphero-cylindrical Lens ($+ s^2 \supset - c^1$ axis 90°)—Periscopic Form.

The final rays from the outermost points, e^0, in the horizontal plane being directed to the focal point F^2, it is evident that the focal line $l^1\ F^1\ l^1$ must become subject to the influence of the spherical refraction in this plane, thereby establishing the points d^1 and d^1, and restricting the magnitude of the focal line to $d^1\ F^1\ d^1$.

The final rays from the outermost points, e^1, in the vertical plane, which in absence of the cylinder would have been directed to the focal point F^2, now cross each other at F^1. Their extremities are therefore displaced from F^2 to d^2 and d^2, thus resulting in the destruction of the focal point F^2, and establishing a limitation of the rays to a *created* focal line $d^2\ F^2\ d^2$.

§ 30. The convex sphero-cylindrical lens of periscopic form, Fig. 26, is constructed by combining a *weaker* concave cylinder, c^1, with a convex spherical surface, s^2, the axis of the cylinder here being placed in the vertical instead of the horizontal plane for the purpose of future reference.

In this case we have given to the spherical surface s^2 a curvature corresponding to the focal point F^1, and to the cylindrical surface c^1, a curvature, which, acting in combination with its associated horizontal meridian of the spherical surface, causes the rays to unite at the focal line $d^2 F^2 d^2$. The reasons given for the destruction of the focal point F^2, in the lens Fig. 25, may in this instance be similarly applied to explain the creation of the primary focal line $d^1 F^1 d^1$, as well as the limitation of the secondary focal line to the magnitude $d^2 F^2 d^2$.

§ 31. A characteristic difference between the double and the periscopic form of astigmatic lens consists in the fact that the positions of the focal lines are interchanged with respect to their correlative elements of creation. Thus, in Fig. 25 the focal line $d^2 F^2 d^2$ corresponds to the initial effect of the spherical surface; whereas, in Fig. 26 the primary focal line $d^1 F^1 d^1$ corresponds to the same.

§ 32. This difference, however, is not material, as it is evident that the magnitude of the focal lines and their distance from the lens are dependent upon the refraction ascribed to its two principal sections; and, since any two given points (d^1 and F^2, F^1 and d^2, m^1 and m^2) definitely fix the position of a line or ray in space, it is further obvious that the direction of all final rays will be identical for any lens* in which the right-angled coördinate meridians of greatest and least refraction are allotted the same.

§ 33. To demonstrate the analysis of formulæ for these equivalents, we shall, in the respective figures, designate the refraction as being expressed by

Ia.	+ 3.5 cyl. axis 180°	⊂⊃ + 1.5 cyl. axis 90°.	(Fig. 24.)
IIa.	+ 1.5 spherical	⊂⊃ + 2 cyl. axis 180°.	(Fig. 25.)
IIIa.	+ 3.5 spherical	⊂⊃ — 2 cyl. axis 90°.	(Fig. 26.)

It being necessary to become thoroughly familiar with the meridians of greatest and least refraction, it is considered expedient to picture these in their respective planes of activity, V and H, as shown in their correlative

* Wherein the rays are incident in the immediate vicinity of the optical axis, and the thickness of the lens is a vanishing quantity in proportion to the focal distances of the surfaces.

sectional diagrams, Fig. 24a, Fig. 25a, Fig. 26a, and to refer to them, as follows:

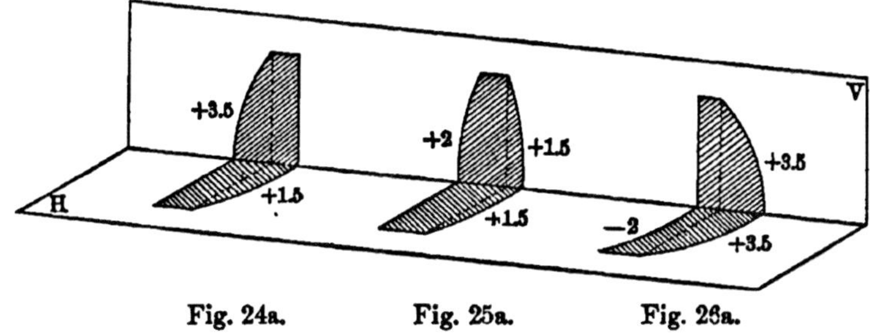

Fig. 24a. Fig. 25a. Fig. 26a.

Formula Ia. + 3.5 cyl. axis 180° ◯ + 1.5 cyl. axis 90°.

Refraction: Fig. 24a. } + 3.5 vertical ◯ + 1.5 horizontal = + 3.5V ◯ + 1.5H.

Formula IIa. + 1.5 spherical ◯ + 2 cyl. axis 180°.

Refraction: Fig. 25a. } + 2 + 1.5 vertical ◯ + 1.5 horizontal = + 3.5V ◯ + 1.5H.

Formula IIIa. + 3.5 spherical ◯ — 2 cyl. axis 90°.

Refraction: Fig. 26a. } + 3.5 vertical ◯ — 2 + 3.5 horizontal = + 3.5V ◯ + 1.5H.

Pursuant to § 32 we find that the lenses Ia, IIa, and IIIa are asymmetrically-refracting equivalents.

§ 34. As the preference is generally given to the double form (Formula IIa), and, under certain circumstances, occasionally to the periscopic (Formula IIIa), we here only give the rules applicable for the conversion of the one into the other formula.

To convert the double into the periscopic form:

Rule 1. **Place the sum of both numerals of refraction as the numeral for the newly-created spherical element,* and combine with the same cylindrical element having its sign and axis reversed.**

---

*The sign of the original spherical remaining unchanged.

To convert the periscopic into the double form:

Rule 2. **Place the difference of both numerals of refraction as the numeral for the newly-created spherical element,* and combine with the same cylindrical element having its sign and axis reversed.**

§ 35. As these lenses are, practically, only used for the correction of anomalies of ocular refraction, it is customary when adapting them to note the positions of the cylindrical axes, which are precisely indicated by the graduations of the trial-frame. This, however, does not change the inherent properties of the lenses, whose meridians of greatest and least refraction are always 90° apart for all possible axial positions between 0° and 180°.

Thus, in the instance of the formula:

$$+ 1.5 \text{ sph.} \bigcirc + 0.50 \text{ cyl. axis } 130^\circ$$

the periscopic form would be expressed according to Rule 1, § 34, by

$$+ 2 \text{ sph.} \bigcirc - 0.50 \text{ cyl. axis } 40^\circ.$$

Inversely, the former may be made the result of the latter by application of Rule 2, § 34.

A table showing the available combinations by crossed convex cylinders, from 0.25D. to 3.5D., is hereto appended, wherein, according to § 24, crossed convex cylinders of identical curvature are substituted by their spherical equivalents.

The diagonal column of spherical lenses divides the table into two sets of compound lenses which are duplicates in refraction, though differing in the positions of their cylindrical axes by 90°.

Thus all lenses in the vertical columns ***beneath*** the spherical are correlative duplicates of the lenses in the horizontal columns to the ***right*** of the same spherical. $(A^1 = a^1)$, $(A^2 = a^2)$, $(A^3 = a^3)$, $(B^1 = b^1)$, $(B^2 = b^2)$, $(B^3 = b^3)$, etc.

* The sign of the original spherical remaining unchanged.

COMPOUND LENSES.

2. CONGENERIC MERIDIANS (CONCAVE).

§ 36. The preceding general principles are alike applicable to the similarly* planned concave compound lenses Figs. 27, 28, 29, in each of which the focal lines, and consequently also the focal interval and circle of least confusion are virtual, and in the negative region before the lens.

All parallel rays incident upon and within the periphery of the circle C, in any of the figures, will, therefore, result in final rays behind the lens which appear to emanate from correlatively established virtual points d^1 and F^2, F^1 and d^2, m^1 and m^2, of and within the limits of the focal lines before the lens. For these lenses, respectively, we allot the refraction as follows:

Ib. — 1.5 cyl. axis 180° ⊃ — 3.5 cyl. axis 90°. (Fig. 27.)

IIb. — 1.5 spherical ⊃ — 2 cyl. axis 90°. (Fig. 28.)

IIIb. — 3.5 spherical ⊃ + 2 cyl. axis 180°. (Fig. 29.)

and which, by a similar method of analysis to § 33, pursuant to § 32, will be found to be asymmetrically-refracting equivalents.

According to Rule 1, § 34, as an instance, the concave sphero-cylindrical lens

— 1.25 sph. ⊃ — 0.75 cyl. axis 160°

may be converted into the periscopic form

— 2 sph. ⊃ + 0.75 cyl. axis 70°,

and *vice versa*, according to Rule 2, § 34.

*The meridian of greatest refraction is here placed in the horizontal instead of the vertical plane.

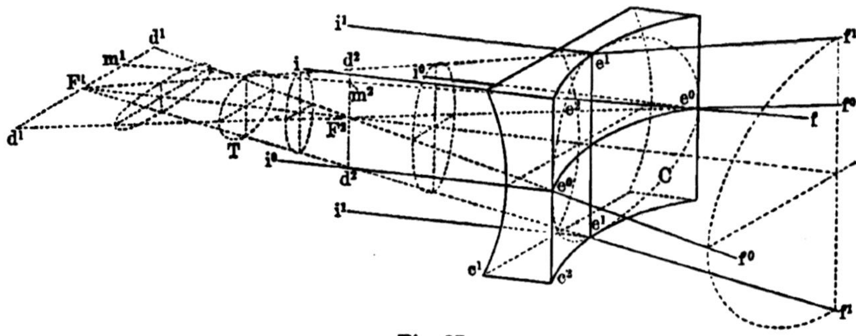

Fig. 27.

Concave Cylindro-cylindrical Lens (— c^1 axis 180° ⊃ — c^2 axis 90°).

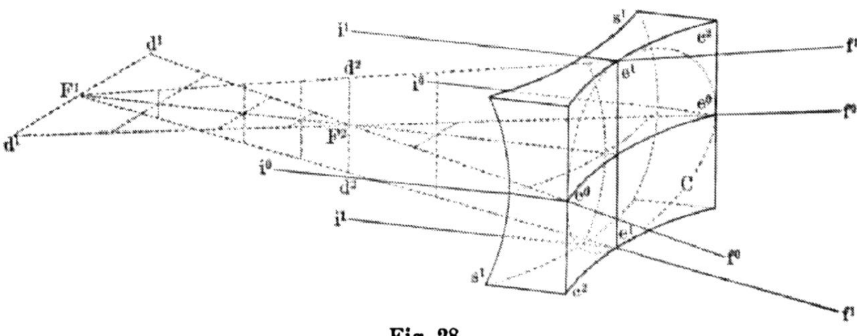

Fig. 28.

Concave Sphero-cylindrical Lens (— s^1 ⊃ — c^2 axis 90°)—Double Form.

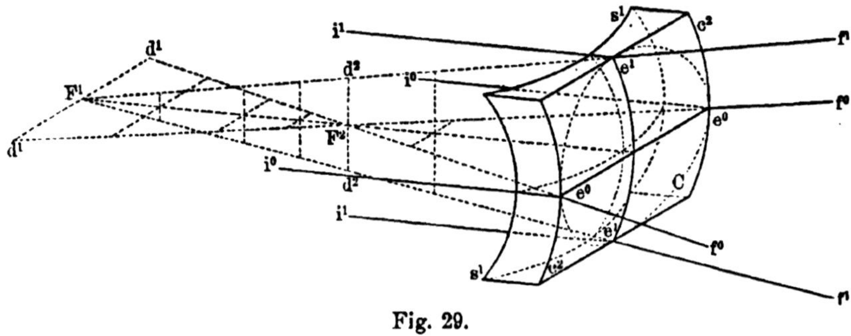

Fig. 29.

Concave Sphero-cylindrical Lens (— s^1 ⊃ + c^2 axis 180°)—Periscopic Form.

In the above figures, i^0, e^0, and f^0 are associated with horizontal, and i^1, e^1, and f^1 with vertical refraction.

COMPOUND LENSES.

3. CONTRA-GENERIC MERIDIANS (CONVEX AND CONCAVE).

§ 37. Hitherto we have considered different amounts of refraction, restricted to the same type, convex or concave, for the principle right-angled

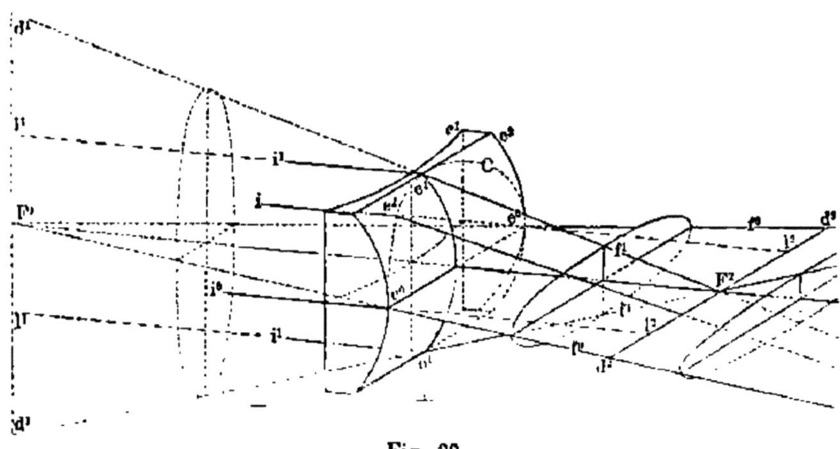

Fig. 30.

Concavo-convex Cylindro-cylindrical Lens (— c^1 axis 90° $\supset$ + c^2 axis 180°).

sections. In contradistinction thereto, and as a final complication, we may combine in a lens different or even like degrees of refraction, though of opposite type; namely, convex in one and concave in the other diametrically-opposed coördinate meridian. As an instance, we may select the compound lens Fig. 30, represented as consisting of a plano-concave, c^1, and a plano-convex cylinder, c^2, so combined as to place their active meridians at right angles to each other.

Independently considered, each cylinder c^1 and c^2 would have its focal line l^1 F^1 l^1, and l^2 F^2 l^2, of original magnitude in the region of its sign — and + respectively, and consequently on opposite sides of the lens.

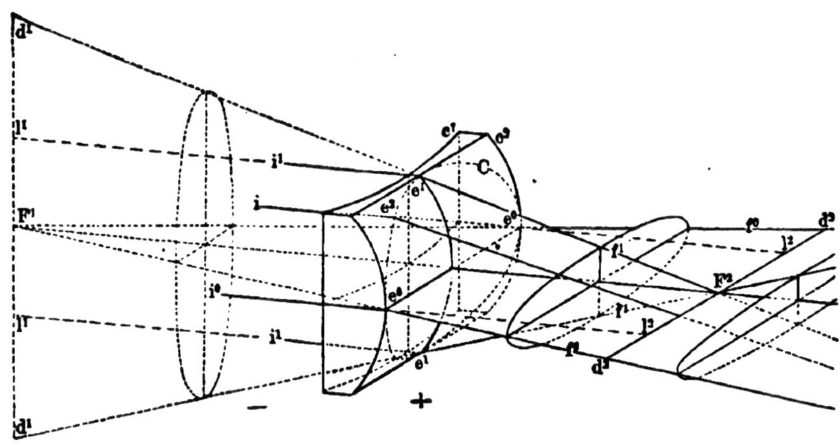

Fig. 30.

Concavo-convex Cylindro-cylindrical Lens (— c^1 axis 90° ⊃ + c^2 axis 180°).

When the cylinders are associated, however, the final rays, which would have been restricted to the limits of the focal line l^2 F^2 l^2 for the cylinder c^2, will, by virtue of the dispersive effect of the cylinder c^1 in the horizontal plane, be confined to an augmented focal line d^2 F^2 d^2, within the limits d^2–d^2, for the outermost rays emanating from the point F^1 of the virtual focal line, l^1 F^1 l^1.

By a similar method of reasoning to § 26, all final rays within the limits of the circle C will be accorded associated vertical and horizontal refraction, culminating in their united intersection of a line d^2 F^2 d^2, of the horizontal plane in the positive region behind the lens. Interception of these rays, by successive transverse vertical planes, will make manifest a demonstration of similarly arranged ellipses, respecting their greatest and least diameters, before and behind the focal line d^2 F^2 d^2. By projecting the final rays into the region of their apparent emanation from before the lens, we would obtain a similar increase of the virtual focal line l^1 F^1 l^1 to the magnitude d^1 F^1 d^1, and to a reversal of the so-defined ellipses, respecting their greatest and least diameters, as shown by the dotted lines in the negative region (Fig. 30).

§ 38. Identical refraction is also preferably obtained in this instance by combinations of spherical and cylindrical surfaces.

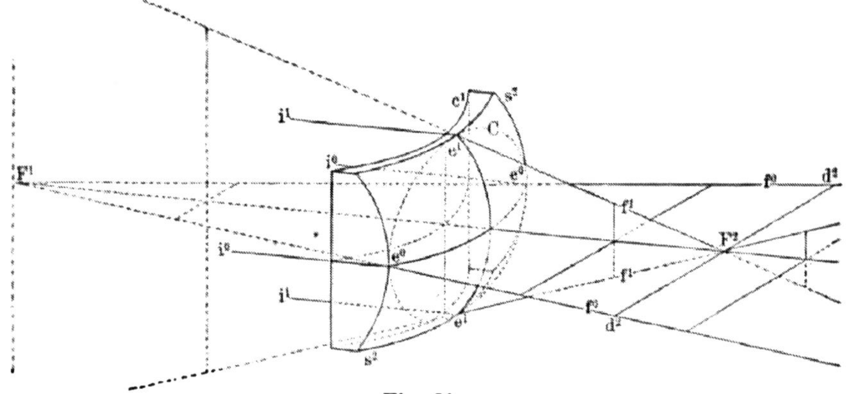

Fig. 31.
Concavo-convex sphero-cylindrical Lens ($+ s^2 \supset - c^1$ axis 90°).

The combination of a convex spherical surface with the active meridian of a stronger concave cylinder creates a periscopic section which is concave; whereas the combination of a concave spherical surface with the active meridian of a stronger convex cylinder results in a periscopic section which is convex. The identity of the refraction for these combinations becomes apparent by reference to the concavo-convex sphero-cylindrical lenses Figs. 31 and 32, in which, by a judicious selection of the respective spherical

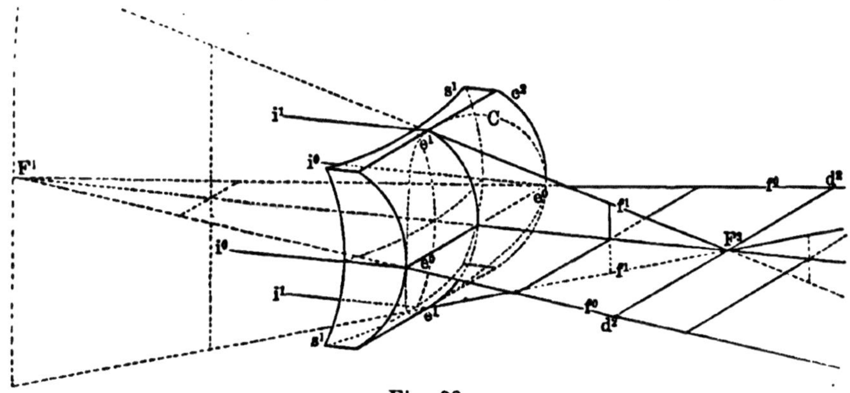

Fig. 32.
Concavo-convex Sphero-cylindrical Lens ($- s^1 \supset + c^2$ axis 180°).

and cylindrical curvatures, according to § 32, the demanded positive and negative elements of refraction for the principal meridians of the crossed cylindrical lens, Fig. 30, are fulfilled.

To illustrate the equality of formulæ characterizing these equivalents, we refer to their correlative sectional diagrams Figs. 30c, 31c, 32c, in the order following:

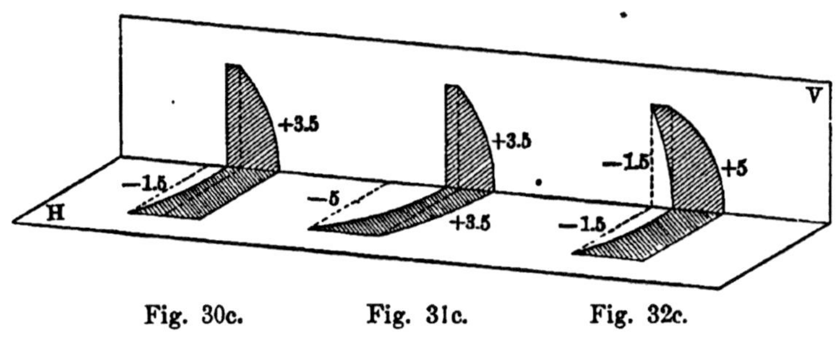

Fig. 30c. Fig. 31c. Fig. 32c.

Formula Ic. — 1.5 cyl. axis 90° ◯ + 3.5 cyl. axis 180°. (Fig. 30.)

Refraction: Fig. 30c. } — 1.5 horizontal ◯ + 3.5 vertical = — 1.5H ◯ + 3.5V.

Formula IIc. + 3.5 spherical ◯ — 5 cyl. axis 90°. (Fig. 31.)

Refraction: Fig. 31c. } — 5 + 3.5 horizontal ◯ + 3.5 vertical = — 1.5H ◯ + 3.5V.

Formula IIIc. — 1.5 spherical ◯ + 5 cyl. axis 180°. (Fig. 32.)

Refraction: Fig. 32c } — 1.5 horizontal ◯ — 1.5 + 5 vertical = — 1.5H ◯ + 3.5V.

§ 39. These lenses being equivalents (see § 32), we here give the only rule required, for reasons later given (§ 40), for converting the cylindro-cylindrical lens (Formula Ic) into the concavo-convex sphero-cylindrical lenses (Formulæ IIc and IIIc).

Rule 3. **Place the sum of both numerals of refraction as the numeral of the newly-created cylindrical element, giving to it both the sign and axis of either cylinder, and combine with the neglected cylindrical numeral and its associated sign as spherical.**

Comparison of the periscopic lenses Figs. 26 and 29 with the lenses Figs. 31 and 32, respectively, exhibits a striking similarity in construction. The characteristic difference between them is that in the latter the cylindrical exceeds the spherical refraction, whereas in the former the reverse is the case.

§ 40. In a case of mixed (contra-generic) astigmatism, demanding the foregoing correction, it becomes necessary to determine the chief meridians — 1.5 and + 3.5 independently of each other, thereby obtaining the combination expressed by Formula Ic, as by an endeavor to correct through introducing a *spherical* element in any proportion or wholly of either equivalent (Formula IIc or IIIc), an improvement in one meridian would always be attended by a proportionate derangement in the other, with a probability of the patient failing to appreciate the benefits of its application.

It is only in consequence of this fact that the lenses of the Formulæ IIc and IIIc are rarely the direct result in subjective optometry, whereas, in cases of regular astigmatism with congeneric meridians, the lenses IIa, IIIa and IIb, IIIb are most apt to be.

§ 41. Astigmatism has, in the main, been attributed to asymmetry of the cornea, though the crystalline lens is often found to be implicated; yet, *specifically*, in a case of mixed astigmatism, *in which the crystalline lens does not assist*, it is improbable that the corneal surface can ever be of the form requisite to include reversed curvatures, Fig. 36. In such instance the ametropia is rather more apt to be one in which an opposite type of astigmatism is in excess of an existing hypermetropia or myopia, respectively. Accepting this to be the case, such an eye would fall heir to the features accredited to hypermetropia or myopia respecting the "nodal points" and "amplitude of accommodation;" wherefore, in prescribing either of the aforesaid sphero-cylindrical equivalents, a preference might be given to that form which would be commensurate with the inherent physical and physiological developments above alluded to.

TORIC LENSES.

CONGENERIC AND CONTRA-GENERIC MERIDIANS.

§ 42. The properties of astigmatic refraction are also fulfilled in a lens by creating for it, opposite to its plane side, a single surface whose diametrically opposed principal meridians are of unequal refraction.

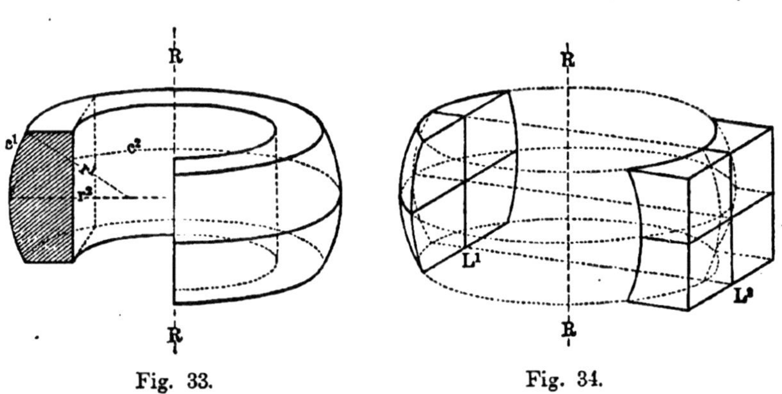

Fig. 33. Fig. 34.

Such a surface, called a torus, is shown in Fig. 33, wherein the curvature c^1 of the radius r^1 and refraction 3D. is rotated upon a vertical axis R so as to create the curvature c^2, whose radius r^2 is chosen to produce 2D.

In Fig. 34 two lenses are shown to be included within the surface so developed and an opposite plane side, the one being a plano-convex toric lens L^1,—the other a plano-concave toric lens L^2.

From the construction it follows that these lenses are each possessed of 3D. of refraction in the vertical, and 2D. of refraction in the horizontal meridian, so that the formulæ for the same may be expressed by

(A_1) [+ 3D. Ref. 90° ⊃ + 2D. Ref. 180°] Tor...............(L^1)

(B_1) [— 3D. Ref. 90° ⊃ — 2D. Ref. 180°] Tor...............(L^2)

as a distinction to the correlative formulæ A_2 and B_2 for a pair of crossed cylinders of identical refraction,

(A_2) + 3 cyl. axis 180° ⊃ + 2 cyl. axis 90°

(B_2) — 3 cyl. axis 180° ⊃ — 2 cyl. axis 90°

and their sphero-cylindrical equivalents, respectively:

A_3) { + 2 sph. ⊃ + 1 cyl. axis 180° (Double Form).
+ 3 sph. ⊃ — 1 cyl. axis 90° (Periscopic Form).

(B_3) { — 2 sph. ⊃ — 1 cyl. axis 180° (Double Form).
— 3 sph. ⊃ + 1 cyl. axis 90° (Periscopic Form).

§ 43. The rotary body shown in Fig. 33 may also be considered to have been created by bending a simple cylindrical lens c^1 to the radius r^2.

In such an attempt, the lens of the Formula A_1 might be obtained by bending a 3D. cylindrical lens to that radius, which effects a refraction of 2D., or a 2D. cylindrical lens to a radius producing a refraction of 3D. In the latter case the lens would merely require to be turned 90° so as to correspond with the rest of the formulæ. The inner or back surface would naturally also require to be restored to a plane, as indicated by the dotted parallelogram in Fig. 33.

The suggested method being impracticable, the process of grinding must be resorted to; although this at present involves more complicated apparatus. A lens having one surface spherical and the other toric is called a sphero-toric lens. Its advantages are explained in a subsequent paper.

§ 44. A toric surface with contra-generic meridians is shown in Fig. 35, wherein the concave section c^1, of the refraction — 3D., is rotated upon the vertical axis R, so as to create the convex curvature c^2, whose refraction is + 2D.

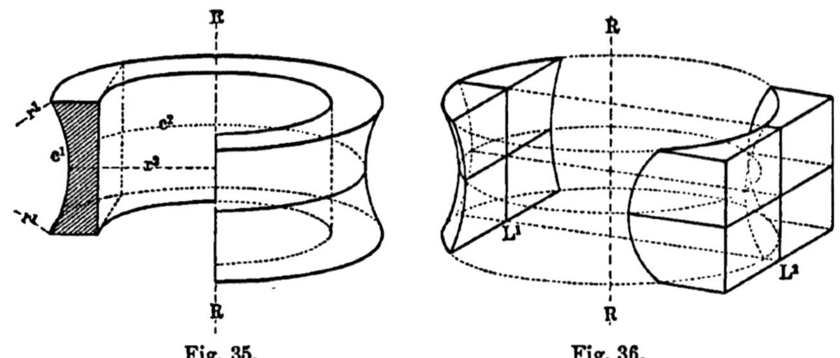

Fig. 35. Fig. 36.

In Fig. 36 two plano-toric lenses L^1 and L^2 are shown to have the same toric surface as in Fig. 35. As the principal meridians in each of these lenses are convex and concave, we may write their formulæ as follows:

(C_1) [- 3D. Ref. 90° ⊃ + 2D. Ref. 180°] Tor................(L^1)

(D_1) [+ 3D. Ref. 90° ⊃ — 2D. Ref. 180°] Tor................(L^2)

so as to distinguish them from the correlative formulæ for crossed cylinders of identical refraction:

(C_2) — 3 cyl. axis 180° ⊃ + 2 cyl. axis 90°

(D_2) + 3 cyl. axis 180° ⊃ — 2 cyl. axis 90°

and their sphero-cylindrical equivalents, which are respectively:

(C_3) { + 2 sph. ⊃ — 5 cyl. axis 180°.
— 3 sph. ⊃ + 5 cyl. axis 90°. }

(D_3) { — 2 sph. ⊃ + 5 cyl. axis 180°.
+ 3 sph. ⊃ — 5 cyl. axis 90°. }

The student should practise transforming optionally chosen formulæ, by applying the rules given in § 34 and § 39, when he may refer to the appended tables to verify the correctness of his own work.

NUMERALS OF REFRACTION.

Focal Distances. Centimeters.	Metric System. Dioptries.	=	Inch System. Approximates.	Focal Distances. U. S. Standard Inches.
400.	0.25		1:160	$157\frac{1}{2}$
200.	0.50		1:80	$78\frac{3}{4}$
133.3	0.75		1:53	$52\frac{1}{2}$
100.	1.		1:40	$39\frac{3}{8}$
80.	1.25		1:32	$31\frac{1}{2}$
66.7	1.50		1:26	$26\frac{1}{4}$
57.1	1.75		1:22	$22\frac{1}{2}$
50.	2.		1:20	$19\frac{11}{16}$
44.4	2.25		1:18	$17\frac{1}{2}$
40.	2.50		1:16	$15\frac{3}{4}$
36.4	2.75		1:14	$14\frac{5}{16}$
33.3	3.		1:13	$13\frac{1}{8}$
30.8	3.25		1:12	$12\frac{1}{8}$
28.6	3.50		1:11	$11\frac{1}{4}$
25.	4.		1:10	$9\frac{7}{8}$
22.2	4.50		1:9	$8\frac{3}{4}$
20.	5.		1:8	$7\frac{7}{8}$
18.2	5.50		1:7	$7\frac{1}{8}$
16.7	6.		$1:6\frac{1}{2}$	$6\frac{9}{16}$
15.4	6.50		1:6	6
14.3	7.		$1:5\frac{1}{2}$	$5\frac{5}{8}$
12.5	8.		1:5	$4\frac{15}{16}$
11.1	9.		$1:4\frac{1}{2}$	$4\frac{3}{8}$
10.	10.		1:4	$3\frac{15}{16}$
9.1	11.		$1:3\frac{1}{2}$	$3\frac{9}{16}$
8.3	12.		$1:3\frac{1}{4}$	$3\frac{9}{32}$
7.7	13.		1:3	$3\frac{1}{32}$
7.1	14.		$1:2\frac{3}{4}$	$2\frac{13}{16}$
6.7	15.			$2\frac{5}{8}$
6.3	16.		$1:2\frac{1}{2}$	$2\frac{7}{16}$
5.5	18.		$1:2\frac{1}{4}$	$2\frac{3}{16}$
5.	20.		1:2	$1\frac{31}{32}$
2.5	40.		1:1	$\frac{63}{64}$

The above table has been arranged for comparison of the metric with the old system of numbering, in which 1 inch was adopted as the unit. A lens of 10, 20, or 40 inches focus is therefore represented as being $\frac{1}{10}$, $\frac{1}{20}$, or $\frac{1}{40}$ of the refraction of the old standard.

The focal distances have been calculated upon the basis that 1 meter = 100 centimeters = 39.37 U. S. standard inches, through dividing each of these equivalents by the dioptral numerals. To render a harmony of the numerals of the two systems possible, it is found necessary to neglect slight fractional variations, as shown in the differences between the divisors in the 3rd with the figures of the 4th column. 1 dioptry being placed as equivalent to $\frac{1}{40}$, lenses of 2, 3, or 4 dioptries may be calculated as $\frac{2}{40} = \frac{1}{20}$, $\frac{3}{40} = \frac{1}{13}$, or $\frac{4}{40} = \frac{1}{10}$, respectively, without materially conflicting with the practical demands upon accuracy in a substitution of one system of numerals for the other.

DIOPTRIES
+0.25 C.90°
+0.50 C.90°
+0.75 C.90°
+1.00 C.90°
+1.25 C.90°
+1.50 C.90°
+1.75 C.90°
+2.00 C.90°
+2.25 C.90°
+2.50 C.90°
+2.75 C.90°
+3.00 C.90°
+3.25 C.90°
+3.50 C.90°

In the a
de
With the
cy

DIOPTRIES
−0.25 C.90°
−0.50 C.90°
−0.75 C.90°
−1.00 C.90°
−1.25 C.90°
−1.50 C.90°
−1.75 C.90°
−2.00 C.90°
−2.25 C.90°
−2.50 C.90°
−2.75 C.90°
−3.00 C.90°
−3.25 C.90°
−3.50 C.90°

In the abo
the
Practical e
For crosse

SECTION II

DIOPTRIC FORMULÆ FOR COMBINED CYLINDRICAL LENSES

APPLICABLE FOR

ALL ANGULAR DEVIATIONS OF THEIR AXES

WITH SIX ORIGINAL DIAGRAMS AND ONE HALF-TONE PLATE

DIOPTRIC FORMULÆ

FOR COMBINED

CONGENERIC CYLINDERS

PLATE I

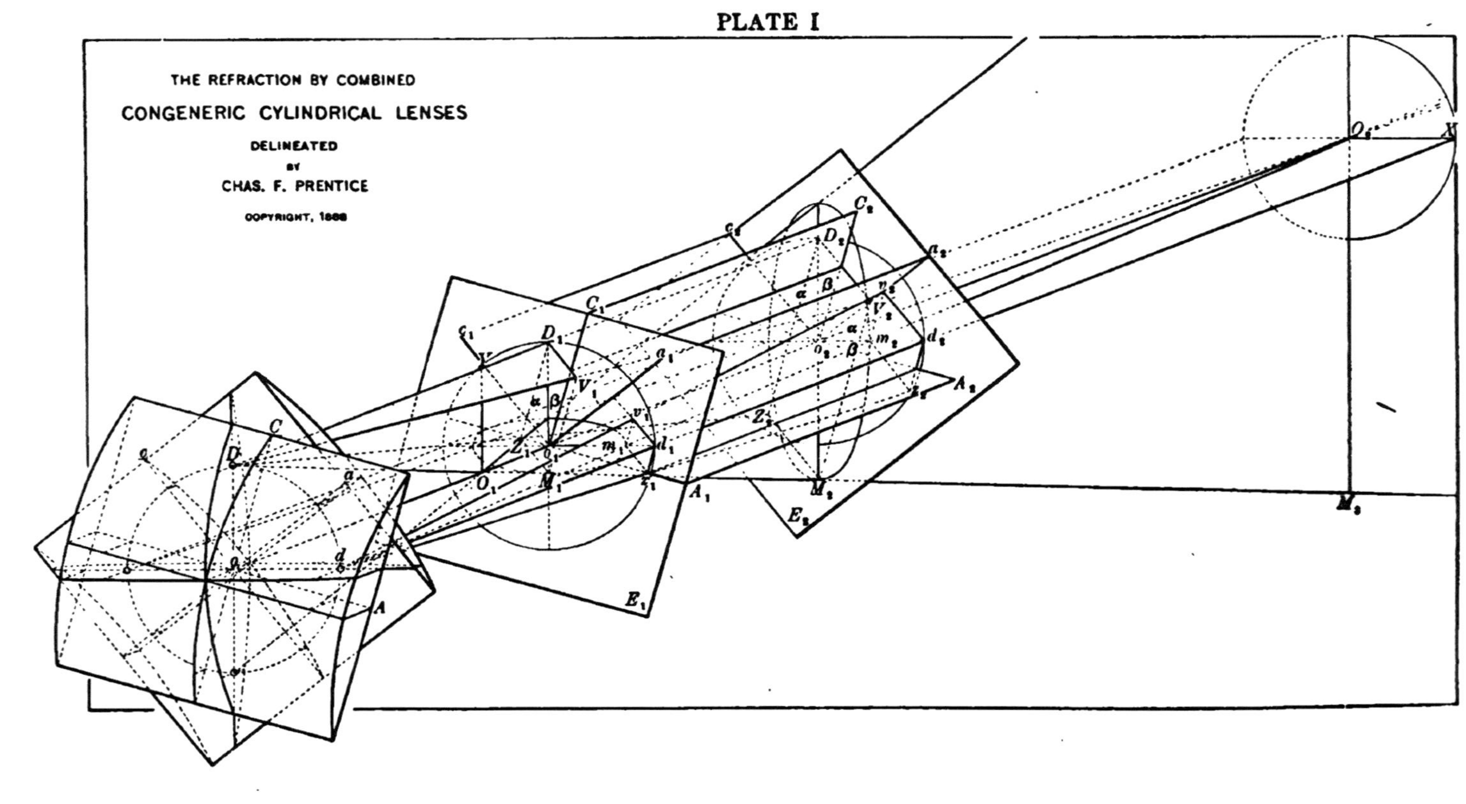
THE REFRACTION BY COMBINED
CONGENERIC CYLINDRICAL LENSES
DELINEATED
BY
CHAS. F. PRENTICE
COPYRIGHT, 1888

I. DIOPTRIC FORMULÆ

FOR COMBINED

CONGENERIC CYLINDRICAL LENSES.

1. RELATIVE POSITIONS OF THE PRIMARY AND SECONDARY PLANES OF REFRACTION.

In the following theorems, a prior knowledge of the established mathematical deductions applied to lenses for parallel rays incident in the immediate vicinity of the optical axis, and in which the lens-thicknesses are considered vanishing quantities in proportion to the focal distances is taken for granted. The formulæ here advanced are therefore dependent upon those which have not been carried beyond *first approximations*. Practically, in almost all cases that occur, the thicknesses of the combined lenses are very small quantities compared to the other dimensions involved, so that we shall consider the cylinders to be so thin that their centers actually coincide, and in which case the focal distances are to be counted from a plane perpendicular to the optical axis, in the optical center of the combined lenses.

In Plate I, two combined convex cylindrical lenses are shown, which, though somewhat at variance with the prescribed conditions of thickness, will, however, better serve to make our subject clear.

The dotted circle shown within the lenses, with its center at the optical center o, shall represent the plane above alluded to.

The *passive* or axial planes of the cylinders are shown by dotted parallelograms at A and a, bisecting each other under the angle $Aoa = \gamma$ in the optical axis at o; and their *active* planes of refraction C and c, which are of necessity at right angles to their correlative axial planes, similarly bisect each other at the same point. Hence, $\sphericalangle\ Coc = \sphericalangle\ Aoa = \gamma$.

The compound lens thus presented consists of two congeneric cylindrical elements, each of which, *independently considered*, will have its corresponding focal plane, which, for convenience, we may term an *elementary* focal plane of the combination. Thus, E_1 and E_2, at the focal distances f_1 and f_2, are the elementary focal planes for the cylinders C and c, respectively. The cylinder C will consequently have the property of deflecting a ray, incident at D, perpendicularly from D_1, in the plane E_1, to the point Z_1 of the axial plane $A_1\ Z_1$; whereas, the cylinder c will have the property of deflecting a ray incident at the same point, perpendicularly from D_2, in the plane E_2, to the point V_2, of the axial plane $a_2\ o_2$.

The greatest amplitude of deflection for C will therefore be $D_1 Z_1$ in the plane E_1, and for c will be $D_2 V_2$ in the plane E_2. It is further manifest that the refracted ray $DV_1 V_2$, contributed by c only, in attaining its greatest deflection $D_2 V_2$, in the plane E_2, will penetrate the plane E_1 at V_1, and in it present a proportionate deflection $D_1 V_1$.

$D_1\ Z_1$ and $D_1 V_1$, being amplitudes of deflection *reduced to the same plane* E_1, will then bear the same relation to each other as their corresponding refractions. Thus

$$D_1 Z_1 : \frac{1}{f_1} = D_1 V_1 : \frac{1}{f_2};$$

or,

$$D_1 Z_1 = \frac{1}{f_1}, \text{ when } D_1 V_1 = \frac{1}{f_2},$$

and which may easily be shown to be the case when the deflections are measured in a plane one inch from the lens.*

Provided, therefore, that the deflections are measured, within the same plane, from a point D_1 of the same line of incidence DD_1, we may determine the resultant of two deflections $D_1 Z_1$ and $D_1 V_1$, for any angular deviation existing between them at D_1, by the physical law governing similarly united forces. $D_1 M_1$, as the diagonal of the parallelogram $D_1 V_1 M_1 Z_1$, will consequently be the resultant deflection accruing from a combination of the cylinders C and c.

* Refraction and Accommodation of the Eye, by E. Landolt, M.D., Paris, translated by C. M. Culver, M.A., M.D., Philadelphia, 1886 (see page 58).

As each cylinder contributes a plane of active and one of passive refraction, we shall evidently obtain two resultant principal planes for their combination, the one of greatest refraction, commonly called the ***primary*** plane, DD_1o_1o, intersecting the angle $Coc = \gamma$ between the active planes of refraction C and c, and one of least refraction, termed the ***secondary*** plane, dd_2o_2o, intersecting the angle $Aoa = \gamma$ between the passive or axial planes A and a.

The primary plane, in penetrating the plane E_1, will consequently divide the angle $C_1o_1c_1 = Coc = \gamma$ into $D_1o_1c_1 = \alpha$ and $D_1o_1C_1 = \beta$. In the plane E_1 we shall then find the angles α and β to be directly dependent upon the associated deflections D_1Z_1 and D_1V_1 for the point D_1. In the plane E_2 a similar division of the angle $A_2o_2a_2$, by the secondary plane, will be rendered dependent upon d_2v_2 and d_2z_2 for the point d_2. As to this, the diagram is believed to be sufficiently clear, without further reference.

Since the resultants D_1M_1 and d_2m_2 define the directions of the refracted rays DM_1 and dm_2, it is further evident that for D and d to be points of the primary and secondary planes, respectively, they will have to be so chosen that D_1M_1 and d_2m_2 shall be directed *towards* the optical axis oo_1o_2; and, as we shall later learn, this is but one of the restrictions which renders a diagram somewhat difficult of construction. The resultant deflections D_1M_1 and d_2m_2 are therefore shown in the primary plane, *coincident* with D_1o_1, and in the secondary plane *coincident* with d_2o_2, respectively.

For all intermediate points of the circle, the resultant deflections deviate *from* the optical axis. This has been taken advantage of in constructing Dr. Burnett's models, and in determining the directions of twelve refracted rays in each of the figures 2, Plates II and IV.

The position of the primary plane DD_1o_1o, shown as dividing the angle $C_1o_1c_1 = \gamma$ so that

$$\gamma = \alpha + \beta, \quad . \quad . \quad . \quad . \quad . \quad . \quad . \quad . \quad . \quad (1)$$

will then be determined by fixing the relations existing between α and β.

In the plane E_1, from the triangle $D_1Z_1M_1$, we have

$$D_1Z_1 : Z_1M_1 = \sin \measuredangle Z_1M_1D_1 : \sin \measuredangle Z_1D_1M_1,$$

$$\measuredangle Z_1M_1D_1 = \measuredangle D_1o_1c_1 = \alpha,$$

by parallelism of Z_1M_1 and c_1o_1; and, for similar reasons,

$$\sphericalangle Z_1D_1M_1 = \sphericalangle D_1M_1V_1 = D_1o_1C_1 = \beta.$$

$$\therefore \quad D_1Z_1 : Z_1M_1 = \sin\alpha : \sin\beta,$$

$$Z_1M_1 = D_1V_1.$$

$$\therefore \quad D_1Z_1 : D_1V_1 = \sin\alpha : \sin\beta. \quad \ldots \ldots \quad (2)$$

In the oblique plane DD_2V_2 we find

$$D_1V_1 : D_2V_2 = DD_1 : DD_2;$$

or, as DD_1 and DD_2 are the focal distances f_1 and f_2 of the cylinders C and c, respectively,

$$D_1V_1 : D_2V_2 = f_1 : f_2. \quad \ldots \ldots \ldots \quad (3)$$

Multiplying the equations (2) and (3), we obtain

$$\frac{D_1Z_1}{D_2V_2} = \frac{\sin\alpha}{\sin\beta}\frac{f_1}{f_2}. \quad \ldots \ldots \ldots \quad (4)$$

Since D_1o_1 is the radius of the circle indicated, we may, for convenience, ascribe to it the value 1. We shall then have

$$D_1Z_1 = \sin \sphericalangle D_1o_1Z_1,$$

$$\sphericalangle D_1o_1Z_1 = C_1o_1Z_1 - \sphericalangle D_1o_1C_1.$$

$$\therefore \quad \sphericalangle D_1o_1Z_1 = 90° - \beta.$$

$$\therefore \quad D_1Z_1 = \sin(90° - \beta) = \cos\beta. \quad \ldots \quad (5)$$

In the plane E_2 we similarly find

$$D_2V_2 = \sin \sphericalangle D_2o_2V_2,$$

$$\sphericalangle D_2o_2V_2 = \sphericalangle V_2o_2c_2 - \sphericalangle D_2o_2c_2.$$

$$\therefore \quad \sphericalangle D_2o_2V_2 = 90° - \alpha.$$

$$\therefore \quad D_2V_2 = \sin(90° - \alpha) = \cos\alpha. \quad \ldots \quad (6)$$

Substituting the values for $D_1 Z_1$ and $D_2 V_2$ from (5) and (6) in the equation (4), we obtain,

$$\frac{\cos\beta}{\cos\alpha}=\frac{\sin\alpha}{\sin\beta}\frac{f_1}{f_2};$$

or, by multiplying both members of equation by 2 and transposing,

$$2\cos\beta\sin\beta=2\cos\alpha\sin\alpha\frac{f_1}{f_2}$$

$$\therefore \sin 2\beta=\sin 2\alpha\frac{f_1}{f_2}. \quad . \quad . \quad . \quad . \quad . \quad . \quad . \quad . \quad (7)$$

The position of the secondary plane dd_2o_2o, shown as dividing the angle $A_2o_2a_2=\gamma$ into $d_2o_2a_2=\alpha$ and $d_2o_2A_2=\beta$, *provided* d_2o_2 *is perpendicular to* D_2o_2, will be determined by similarly fixing the relations between α and β. Here it can also be shown that

$$d_2z_2 : d_2v_2 = \cos\alpha : \cos\beta. \quad . \quad . \quad . \quad . \quad . \quad . \quad . \quad (8)$$

$$d_1z_1 : d_2z_2 = f_1 : f_2. \quad . \quad . \quad . \quad . \quad . \quad . \quad . \quad . \quad (9)$$

$$d_1z_1 = \sin\beta. \quad . \quad . \quad . \quad . \quad . \quad . \quad . \quad . \quad . \quad (11)$$

$$d_2v_2 = \sin\alpha. \quad . \quad . \quad . \quad . \quad . \quad . \quad . \quad . \quad . \quad (12)$$

whereby, as before, $\sin 2\beta=\sin 2\alpha\frac{f_1}{f_2}$.

We therefore conclude that:

1. The primary and secondary planes of refraction are at right angles to each other for any angular deviation of the axes of two combined congeneric cylindrical lenses.

In a further consideration of the relation (7), $\sin 2\beta=\sin 2\alpha\frac{f_1}{f_2}$, we observe that the sines of the angles 2α and 2β, which are each always less than 90°, merely differ by the co-efficient $\frac{f_1}{f_2}$.

If, therefore, $f_2=f_1$, which is the case when the cylinders are of equal refraction, the $\sin 2\beta$ will be equal to the $\sin 2\alpha$, and which can only be the case when $\alpha=\beta$, or, as $\alpha+\beta=\gamma$, when $\alpha=\beta=\frac{\gamma}{2}$; hence:

2. For combined congeneric cylinders of equal refraction, the primary plane equally divides the angle between the active planes of the cylinders, and the secondary plane similarly divides the angle between the axial planes of the cylinders.

In case, however, $f_2 > f_1$, which is the case when the refraction of the cylinder *C* is greater than *c*, then $\sin 2\alpha > \sin 2\beta$, or, when $\alpha > \beta$, so that

3. For combined congeneric cylinders of unequal refraction, the primary plane, in dividing the angle between the active planes of the cylinders, will be nearer to the active plane of the stronger cylinder, and the secondary plane consequently nearer to the axial plane of the same cylinder.

This is also demonstrated in the diagram.

As, for a combination of two cylinders, *C* and *c*, under given angular deviation of their axes, the only known quantities will be f_1, f_2, and γ, it will be necessary to express α and β in terms of f_1, f_2, and γ.

This is accomplished through the equations (1) and (7):

$$\gamma = \alpha + \beta$$
$$\sin 2\beta = \sin 2\alpha \frac{f_1}{f_2},$$

when, after proper substitution and reduction, we obtain:

$$\cos \alpha = \sqrt{\frac{1}{2} + \frac{1}{2} \frac{f_1 + f_2 \cos 2\gamma}{\sqrt{f_1^2 + 2f_1 f_2 \cos 2\gamma + f_2^2}}} \quad . \quad . \quad . \quad (I)$$

It will be unnecessary to seek β in the same manner, since, through (1), we find $\beta = \gamma - \alpha$.

When reducing this formula, for any given value of γ, pursuant to reasons later given, it should be observed that $f_2 > f_1$, in which case α, within the angle γ, is to be counted from the axis of the *weaker* cylinder.

2. POSITIONS OF THE PRIMARY AND SECONDARY FOCAL PLANES.

As the plane DD_1o_1o is the primary plane, it follows that all parallel rays incident in it between D and o will, after refraction, intersect the optical axis oo_1 at some point, which will be a point of the primary focal line. Therefore the resultant ray DM_1M_2, in attaining its greatest deflection D_1M_1 in the elementary plane E_1, will establish the position of the primary focal line, through its previous intersection of the optical axis oo_1, at the point O_1.

In the secondary plane dd_1o_1o, for similar reasons O_2 will be a point of the secondary focal line, though this point of intersection of the final ray dm_1m_2 with the optical axis is more distant, in consequence of the inferior deflection d_2m_2 in the plane E_2.

Similar resultant deflections, at opposite cardinal points of the circle within the lens, define the directions of their corresponding refracted rays. These rays not only limit the major and minor axes of the ellipses shown in the planes E_1 and E_2, but also determine the lengths of the focal lines at O_1 and O_2. Thus O_2M_2 represents one-half of the secondary focal line at O_2. The primary focal line, in the *secondary* plane, perpendicular to YO_1 at O_1, has been omitted, to avoid possible misinterpretation of more important points of reference in the diagram. All rays parallel to the optical axis, incident at intermediate points of the circle within the lens, will, upon refraction, intersect the planes E_1 and E_2 at correlative points of the ellipses drawn thereon.

The circle of least confusion, T, will lie between the planes E_1 and E_2. (See Plate II, Fig. 2.) Its position may be determined through a simple formula given by Prof. W. Steadman Aldis, of the University College, Auckland, New Zealand, in his discussion of the focal interval resulting from rays obliquely incident upon a spherical lens.*

Our object being to determine the distances of the primary and secondary focal lines, or planes, from the principal plane within the combined cylinders, we shall proceed as follows:

* Elementary Treatise on Geometrical Optics, W. S. Aldis, M.A., Cambridge, 1886 (see page 39).

In the primary plane DD_1M_1, we have

$$DY : DD_1 = YO_1 : D_1M_1.$$

Substituting, $DY = O_1o = F_1$ as the primary focal distance;

$$DD_1 = f_1;$$

$$YO_1 = D_1o_1 = \text{radius} = 1.$$

$$\therefore\ F_1 = \frac{f_1}{D_1M_1} \quad . \quad . \quad . \quad . \quad . \quad . \quad . \quad . \quad . \quad . \quad (26)$$

In the parallelogram $D_1V_1M_1Z_1$, the angle between the forces, D_1V_1 and D_1Z_1, being equal to $\measuredangle\ C_1o_1c_1 = \gamma$, we have, as the resultant deflection,

$$D_1M_1 = \sqrt{(D_1Z_1)^2 + (D_1V_1)^2 + 2\,(D_1V_1)\,(D_1Z_1)\cos\gamma}, \qquad (27)$$

in conformity with the statical formula,

$$R = \sqrt{P^2 + Q^2 + 2PQ\cos\gamma},$$

for forces P and Q, acting at the same point, within the same plane, under the angle γ.

Substituting in (27) the value of $D_1Z_1 = \cos\beta$, from (5); and of $D_1V_1 = \frac{f_1}{f_2}D_2V_2$, from (3), $= \frac{f_1}{f_2}\cos\alpha$, from (6), we obtain,

$$D_1M_1 = \overline{\cos^2\beta + \left(\frac{f_1}{f_2}\right)^2\cos^2\alpha + 2\frac{f_1}{f_2}\cos\alpha\cos\beta\cos\gamma}.$$

Introducing this value for D_1M_1 in (26),

$$F_1 = \frac{f_1}{\sqrt{\cos^2\beta + \left(\frac{f_1}{f_2}\right)^2\cos^2\alpha + 2\frac{f_1}{f_2}\cos\alpha\cos\beta\cos\gamma}} \quad . \quad . \quad . \quad (28)$$

By substituting the proper values for α and β, from equation (1), after adequate reduction we obtain

$$F_1 = \frac{f_1f_2}{\sqrt{\tfrac{1}{2}\left[f_1(f_1 + f_2\cos 2\gamma) + (f_1 + f_2)\left(f_2 + \sqrt{f_1^2 + 2f_1f_2\cos 2\gamma + f_2^2}\right)\right]}}.$$

$$. \quad . \quad . \quad . \quad . \quad . \quad . \quad (30)$$

Transforming, and substituting $1 - 2\sin^2\gamma$ for $\cos 2\gamma$, we may, for convenience in calculating, preferably write

$$F_1 = \frac{f_1 f_2}{\sqrt{\frac{(f_1+f_2)^2}{2} - f_1 f_2 \sin^2\gamma + (f_1+f_2)\sqrt{\frac{(f_1+f_2)^2}{4} - f_1 f_2 \sin^2\gamma}}}. \qquad \text{(II)}$$

When the cylinders are of equal refraction, f_1 being equal to $f_2 = f$, the above assumes the simple form,

$$F_1 = \frac{f}{1 + \cos\gamma}. \quad . \quad . \quad . \quad . \quad . \quad . \quad . \quad . \quad . \qquad \text{(IV)}$$

In the secondary plane dd_2XO_2, we have

$$dX : dd_2 = XO_2 : d_2m_2.$$

Substituting, $dX = O_2o = F_2$ as the secondary focal distance;

$$dd_2 = f_2;$$

$$XO_2 = \text{radius} = 1.$$

$$\therefore\ F_2 = \frac{f_2}{d_2m_2}. \quad . \quad . \quad . \quad . \quad . \quad . \quad . \quad . \quad . \qquad (31)$$

In the parallelogram $d_2v_2m_2z_2$, the angle between the forces, d_2v_2 and d_2z_2, being equal to $\sphericalangle\, v_2d_2z_2 = 180° - \sphericalangle\, A_2o_2a_2 = 180° - \gamma$,

$$\therefore\ d_2m_2 = \sqrt{(d_2z_2)^2 + (d_2v_2)^2 + 2\,(d_2v_2)\,(d_2z_2)\cos(180° - \gamma)}.$$

Substituting the value for $d_2z_2 = \frac{f_2}{f_1}\, d_1z_1$, from (9), $= \frac{f_2}{f_1} \sin\beta$, from (11); and for $d_2v_2 = \sin\alpha$, from (12), we obtain,

$$d_2m_2 = \sqrt{\left(\frac{f_2}{f_1}\right)^2 \sin^2\beta + \sin^2\alpha - 2\,\frac{f_2}{f_1} \sin\alpha \sin\beta \cos\gamma};$$

which, introduced in (31) and being multiplied in the numerator and denominator by $\frac{f_1}{f_2}$, gives

$$F_2 = \frac{f_1}{\sqrt{\sin^2\beta + \left(\frac{f_1}{f_2}\right)^2 \sin^2\alpha - 2\frac{f_1}{f_2}\sin\alpha\sin\beta\cos\gamma}}.$$

Through substitution of the proper values for α and β, from equation (1), after suitable reduction we find

$$F_2 = \frac{f_1 f_2}{\sqrt{\tfrac{1}{2}[f_1(f_1+f_2\cos 2\gamma) + (f_1+f_2)(f_2-\sqrt{f_1^2+2f_1f_2\cos 2\gamma+f_2^2})]}} \quad \ldots\ldots\ldots \quad (33)$$

Substituting, $\cos 2\gamma = 1 - 2\sin^2\gamma$,

$$F_2 = \frac{f_1 f_2}{\sqrt{\frac{(f_1+f_2)^2}{2} - f_1f_2\sin^2\gamma - (f_1+f_2)\sqrt{\frac{(f_1+f_2)^2}{4} - f_1f_2\sin^2\gamma}}} \quad \ldots\ldots\ldots \quad (III)$$

This formula, reduced for cylinders of équal refraction, f_1 being equal to $f_2 = f$, becomes

$$F_2 = \frac{f}{1-\cos\gamma}. \quad \ldots\ldots\ldots \quad (V)$$

It may be of interest to note that these formulæ differ from those given for F_1 merely by the minus sign in the denominator.

The preceding formulæ being alike applicable for combinations of convex or concave cylinders, the foci f_1 and f_2 are to be introduced as positive values, merely with the restriction that f_2 be greater than or equal to f_1, in either case.

3. RELATIONS BETWEEN THE PRIMARY AND SECONDARY FOCAL PLANES.

Since F_1 and F_2 have been shown to be dependent upon f_1, f_2, and γ, it is evident that, for fixed values of f_1 and f_2, the resultant foci will be rendered dependent entirely upon whatever value may be given to the angle γ.

It is further obvious that the refraction of one cylinder will be affected most by the other when their axes coincide, or when $\gamma = 0°$, and least when their axes are at right angles to each other, or when $\gamma = 90°$.

We shall, consequently, fix upon the limits of F_1 and F_2 for these extremes of γ.

Introducing $\gamma = 0°$, and consequently $\cos 2\gamma = +1$, into the formulæ (30) and (33), we obtain, for $f_2 > f_1$,

$$F_1 = \frac{f_1 f_2}{\sqrt{\frac{1}{2}\left[f_1(f_1 + f_2) + (f_1 + f_2)(f_2 + f_1 + f_2)\right]}} = \frac{f_1 f_2}{f_1 + f_2}.$$

$$F_2 = \frac{f_1 f_2}{\sqrt{\frac{1}{2}\left[f_1(f_1 + f_2) + (f_1 + f_2)(f_2 - f_1 - f_2)\right]}} = \frac{f_1 f_2}{0} = \infty.*$$

$$\therefore\ F_1 : F_2 = \frac{f_1 f_2}{f_1 + f_2} : \infty. \quad \ldots \ldots \quad (34)$$

For $F_1 = \frac{f_1 f_2}{f_1 + f_2}$, we shall have as the refraction

$$\frac{1}{F_1} = \frac{1}{f_1} + \frac{1}{f_2};\ \text{consequently,}$$

4. When the axes of the congeneric cylinders coincide, the primary focal plane will correspond to that focal plane which is defined by the sum of the refractions of the cylinders, whereas the secondary focal plane will be at infinity.

This is shown in Plate II, Fig. 1.

Introducing $\gamma = 90°$, and consequently $\cos 2\gamma = \cos 180° = -1$, into (30) and (33), we have, for $f_2 > f_1$,

$$F_1 = \frac{f_1 f_2}{\sqrt{\frac{1}{2}\left[-f_1(f_2 - f_1) + (f_1 + f_2)(f_2 + f_2 - f_1)\right]}} = \frac{f_1 f_2}{f_2} = f_1.$$

$$F_2 = \frac{f_1 f_2}{\sqrt{\frac{1}{2}\left[-f_1(f_2 - f_1) + (f_1 + f_2)(f_2 - f_2 + f_1)\right]}} = \frac{f_1 f_2}{f_1} = f_2.$$

$$\therefore\ F_1 : F_2 = f_1 : f_2. \quad \ldots \ldots \quad (35)$$

* ∞ The sign for infinity.

As f_1 and f_2 correspond to the positions of the elementary planes E_1 and E_2, it follows that

5. The primary and secondary focal planes coincide with their correlative elementary focal planes, when the axes of the congeneric cylinders of unequal refraction are at right angles to each other.

This is demonstrated in Plate II, Fig. 2.

In the same relation (35), if $f_1 = f_2$, then $F_1 = F_2$, or

6. The primary, secondary, and elementary focal planes all merge into one plane, when the axes of the congeneric cylinders of equal refraction are at right angles to each other.

As in this case we have but one focal plane, the refraction corresponds to that of a spherical lens.

F_1 being chosen to signify the primary focal distance, it will have to be less than F_2, yet if $f_1 > f_2$, we should find, in consequence of the relation 35), that $F_1 > F_2$. To retain the significances of F_1 and F_2, it will therefore be necessary to substitute f_2 by the greater given value of cylindrical focus, and f_1 by the lesser, as stated under the formulæ, page 62.

Owing to the previous considerations; between the limits of 0° and 90° for γ, we are then to conclude that F_1 will vary between $\frac{f_1 f_2}{f_1 + f_2}$ and f_1, while F_2 varies between ∞ and f_2, as the nearest and most remote limits of focal distance for F_1 and F_2, respectively.

As an illustration, let Fig. 1, Plate II, represent two combined convex cylinders of unequal refraction, with their axes coincident, and so united as to permit of the rotation of one of the cylinders upon the true planes of their faces, about the optical center o.

In the position shown ($\gamma = 0°$), the shortest possible focal distance F_1 of the primary focal line will be $\frac{f_1 f_2}{f_1 + f_2}$, which corresponds to the combined refraction, $\frac{1}{} + \frac{1}{f_2}$, of the cylinders in the active plane. In the secondary plane, $F_2 = \infty$; consequently, $\frac{1}{F_2} = \frac{1}{\infty} = 0$, which corresponds to the refraction in the axial or passive plane of the cylinders.

The slightest change in the position of one of the cylindrical axes will give rise to a definite value of the angle γ in the Formula III, thereby bringing F_2 within the limits of finite distance, while decreasing the value of F_1 in the Formula II.

For each successive increase in the angle γ, the primary focal plane corresponding to F_1, will recede farther and farther from the combined lenses towards E_1, while the secondary focal plane, corresponding to F_2, approaches nearer and nearer from ∞ to E_2, until $\gamma = 90°$, when F_1 will have reached E_1 on the moment that F_2 merges into E_2, as shown in Plate II, Fig. 2.

Rotation of one of the cylinders is thus associated with corresponding changes in the distances F_1 and F_2, while the movements of their correlative focal planes will be in opposite directions to each other; which proves that:

7. The primary and secondary focal planes are conjugate planes, subject to variations of the angle between the axes of the congeneric cylinders.

In order to comply with this law, in constructing the Plate I, it has been necessary to select elementary foci in marked disproportion to the curvatures of the cylinders; otherwise the secondary focus F_2 could not be brought within the space allotted for the diagram.

DIOPTRIC FORMULÆ

FOR COMBINED

CONTRA-GENERIC CYLINDERS

II. DIOPTRIC FORMULÆ

FOR COMBINED

CONTRA-GENERIC CYLINDERS

---

1. RELATIVE POSITIONS OF THE PRINCIPAL POSITIVE AND NEGATIVE PLANES OF REFRACTION.

In a combination of convex and concave cylinders, we can no longer have the primary and secondary planes, which we have learned to consider as planes of greatest and least refraction, but, instead, we shall have a plane of greatest positive and one of greatest negative refraction, synonymously with the generally-adopted distinction between convex and concave lenses, as designated by the signs + (plus) and — (minus), respectively. As the refractions of the convex and concave elements in the combination are opposing forces, the plane of greatest positive refraction will evidently lie between the active plane of the convex and the axial plane of the concave cylinder, whereas the plane of greatest negative refraction will be between the active plane of the concave and the axial plane of the convex cylinder.

In Plate III, therefore, the plane DD_1o_1o of greatest positive refraction is shown between c and A, and the plane dd_1o_1o of greatest negative refraction between C and a. These planes, being at right angles to each other, divide each of the angles $A_1o_1c_1$ and $C_1o_1a_1$ into α and β.

To establish the formulæ for combined contra-generic cylinders, we shall therefore have to ascribe another significance to the angles α and β.

The deviation of the axes Aoa is equal to angle $A_1o_1a_1 = \gamma$, and, since c_1o_1 is perpendicular to a_1o_1, $\alpha + \beta + \gamma$ is equal to 90°; consequently,

$$\alpha + \beta = 90^\circ - \gamma. \quad . \quad . \quad . \quad . \quad . \quad . \quad (36)$$

The elementary focal planes E_0 and E_1, corresponding to the focal distances f_0 and f_1, respectively, are exhibited on opposite sides of the combined cylinders; since E_0, for the concave cylinder, is virtual and in the negative region before the lens, whereas E_1, for the convex cylinder, is in the positive region behind the lens. Consequently for the point D, the convex cylinder c contributes as its greatest amplitude of deflection D_1Z_1, perpendicular to a_1o_1 in the plane E_1. The greatest amplitude of deflection for the concave cylinder C is D_0V_0, perpendicular to A_0o_0 in the virtual plane E_0. As the incident ray at D will be refracted by the concave cylinder, as if emanating from a correlative point V_0 of the virtual axial line V_0o_0, it is evident that the direction of the ray refracted by it will be V_0DV_1. The proportionate deflection contributed by the concave cylinder, measured in the plane E_1 will consequently be D_1V_1.

Provided the point D is properly chosen, it will be a point of the plane of greatest positive refraction, that is to say, when the resultant deflection D_1M_1, accruing from the associated deflections D_1V_1 and D_1Z_1 in the parallelogram of forces $D_1V_1M_1Z_1$, is directed *towards* the optical axis.

To insure D_1M_1 being so directed, it is obvious that the associated deflections, D_1Z_1 and D_1V_1, must also be measured in the plane E_1, in the positive region behind the lens.

Similar reasoning will apply to the point d as being in the plane dd_1o_1o of greatest negative refraction. In this instance d_1m_1 being a force directed *from* the optical axis, in the plane E_1, is to be taken negative, synonymously with the plane of greatest negative refraction.

The relations between α and β are to be determined by an analogous method to the one given for congeneric cylinders, whereby we obtain

$$\sin 2\alpha = \sin 2\beta \frac{f_1}{f_0}, \quad . \quad . \quad . \quad . \quad . \quad . \quad . \quad (37)$$

as defining the positions of the planes of greatest positive and negative refraction, which are again at right angles to each other.

We here also find the sines of the angles, 2α and 2β, to differ by the coefficient $\frac{f_1}{f_0}$. Hence, when $f_0 = f_1$, we shall have $\alpha = \beta = \frac{90° - \gamma}{2}$, or,

8. For combined contra-generic cylinders of equal refraction, the plane of greatest positive refraction equally divides the angle between the active plane of the convex and the axial plane of the concave cylinder; and the plane of greatest negative refraction similarly divides the angle between the active plane of the concave and the axial plane of the convex cylinder.

In case $f_0 > f_1$, then $\beta > \alpha$; or,

9. When the convex cylinder is stronger than the concave cylinder, the plane of greatest positive refraction will be nearer to the active plane of the convex, while the plane of greatest negative refraction will be proportionately farther from the active plane of the concave cylinder.

In case $f_1 > f_0$, then $\alpha > \beta$; or,

10. When the concave cylinder is stronger than the convex cylinder, the plane of greatest negative refraction will be nearer to the active plane of the concave, while the plane of greatest positive refraction will be proportionately farther from the active plane of the convex cylinder.

This is manifest in the diagram.

The values of α and β may be expressed in terms of f_1, f_0 and γ in a similar manner to that shown in the previous theorem, when it can be shown that,

$$\cos\alpha = \sqrt{\frac{1}{2} + \frac{1}{2}\,\frac{f_0 - f_1 \cos 2\gamma}{\sqrt{f_0^2 - 2f_0 f_1 \cos 2\gamma + f_1^2}}} \quad . \; . \; . \; \text{(VI)}$$

This and the transposed equation (36), $\beta = 90° - (\gamma + \alpha)$, suffice to locate the positions of the principal planes of refraction; the angle α being counted from the axis of the *convex* cylinder.

2. POSITIONS OF THE POSITIVE AND NEGATIVE FOCAL PLANES.

The positions of the positive and negative focal planes will evidently here also be determined by the resultant rays, DM_1 and dm_1, and their correlative intersections with the optical axis at O_1 and O_0.

$O_1\ m_3$ will therefore represent one-half the focal line in the positive region behind the lenses, and O_0M_3 one-half the virtual focal line in the negative region before the same.

The ellipses shown in the planes E_1 and E_0 are of the same significance in this as in the preceding combination.

In the plane of greatest positive refraction, DD_1YO_1, we have

$$DY : DD_1 = YO_1 : D_1M_1.$$

Substituting, $DY = O_1o = F_1$ as the positive focal distance;

$$DD_1 = f_1;$$

$$YO_1 = Do = \text{radius} = 1.$$

$$\therefore\ F_1 = \frac{f_1}{D_1M_1} \quad \ldots\ldots\ldots\ldots \quad (47)$$

In the parallelogram $D_1V_1M_1Z_1$, the angle between the forces, D_1V_1 and D_1Z_1, is equal to $180° - \gamma$, since $D_1Z_1 \perp Z_1o_1$, and $D_1V_1 \perp A_1o_1$.

$$\therefore\ D_1M_1 = \sqrt{(D_1Z_1)^2 + (D_1V_1)^2 + 2\,(D_1Z_1)\,(D_1V_1)\cos(180° - \gamma)}.$$

In the oblique plane $D_0V_0DV_1D_1$, we find,

$$D_1V_1 : DD_1 = D_0V_0 : DD_0.$$

$$D_0V_0 = \sin \measuredangle D_0o_0A_0 = \sin \measuredangle D_1o_1A_1 = \sin \beta.$$

$$DD_0 = f_0.$$

$$\therefore\ D_1V_1 = \frac{f_1}{f_0}\sin\beta.$$

$$D_1Z_1 = \sin(\sphericalangle Z_1o_1c_1 - \sphericalangle D_1o_1c_1) = \sin(90^\circ - \alpha) = \cos\alpha.$$

Substituting these values in the equation for $D_1 M_1$, equation (47) becomes,

$$F_1 = \frac{f_1}{\sqrt{\cos^2\alpha + \left(\frac{f_1}{f_0}\right)^2\sin^2\beta - 2\frac{f_1}{f_0}\sin\beta\cos\alpha\cos\gamma}},$$

and which, through equation (36), may be given the form:

$$F_1 = \frac{f_1 f_0}{\sqrt{\tfrac{1}{2}[f_1(f_1 - f_0\cos 2\gamma) + (f_0 - f_1)(f_0 + \sqrt{f_0^2 - 2f_0 f_1\cos 2\gamma + f_1^2})]}}. \qquad (48)$$

Substituting, $\cos 2\gamma = 1 - 2\sin^2\gamma$,

$$F_1 = \frac{f_1 f_0}{\sqrt{\frac{(f_0 - f_1)^2}{2} + f_0 f_1\sin^2\gamma + (f_0 - f_1)\sqrt{\frac{(f_0 - f_1)^2}{4} + f_0 f_1\sin^2\gamma}}} \qquad \text{(VII)}$$

This formula, when reduced for cylinders of equal positive and negative refraction, f_0 being equal to $f_1 = f$, assumes the simple form

$$F_1 = \frac{f}{\sin\gamma}. \qquad \text{(IX)}$$

In the plane of greatest negative refraction, $d_1m_1dO_0X$, we obtain,

$$dX : dd_1 = XO_0 : d_1m_1.$$

Substituting, $dX = O_0o = -F_0$ as the negative focal distance;

$$dd_1 = f_1;$$

$$XO_0 = do = \text{radius} = 1.$$

$$\therefore\ -F_0 = -\frac{f_1}{d_1m_1}; \qquad (49)$$

since d_1m_1 is to be taken negative.

In the parallelogram $d_1v_1m_1z_1$, the angle between the forces, d_1v_1 and d_1z_1, is again $180° - \gamma$; hence,

$$d_1m_1 = \sqrt{(d_1z_1)^2 + (d_1v_1)^2 + 2\,(d_1z_1)\,(d_1v_1)\cos(180° - \gamma)}.$$

In the oblique plane $d_0v_0dv_1d_1$, we find,

$$d_1v_1 : dd_1 = d_0v_0 : dd_0.$$

$$d_0v_0 = \sin(\sphericalangle D_0\rho_0d_0 - \sphericalangle D_0\rho_0A_0) = \sin(90° - \sphericalangle D_1o_1A_1)$$

$$= \sin(90° - \beta) = \cos\beta.$$

$$dd_0 = f_0.$$

$$\therefore\ d_1v_1 = \frac{f_1}{f_0}\cos\beta.$$

$$d_1z_1 = \sin \sphericalangle d_1o_1z_1 = \sin\alpha.$$

Substituting these values in the equations for d_1m_1 and (49), we have,

$$-F_0 = -\frac{f_1}{\sqrt{\sin^2\alpha + \left(\frac{f_1}{f_0}\right)^2\cos^2\beta - 2\frac{f_1}{f_0}\sin\alpha\cos\beta\cos\gamma}}$$

and which, by the aid of equation (36) may be written, $-F_0 =$

$$-\frac{f_1f_0}{\sqrt{\tfrac{1}{2}\left[f_1(f_1 - f_0\cos 2\gamma) + (f_0 - f_1)(f_0 - \sqrt{f_0^2 - 2f_0f_1\cos 2\gamma + f_1^2})\right]}}. \qquad (50)$$

$\therefore\ -F_0 =$

$$-\frac{f_1f_0}{\sqrt{\frac{(f_0-f_1)^2}{2} + f_0f_1\sin^2\gamma + (f_1 - f_0)\sqrt{\frac{(f_0-f_1)^2}{4} + f_0f_1\sin^2\gamma}}}, \qquad \text{(VIII)}$$

which differs from the formula given for F_1 merely by a transposition of the elements in the factor before the second radical, and, consequently, when reduced to cylinders of equal refraction, also becomes

$$- F_0 = - \frac{f}{\sin \gamma}. \quad . \quad . \quad . \quad . \quad . \quad . \quad . \quad . \quad (X)$$

The formulæ (IX) and (X) correspond to those which were applied to the Stokes Lens.

In reducing the preceding formulæ for given values of cylindrical foci, f_0 is to be substituted by the focus of the concave, and f_1 by the focus of the convex cylinder, both being introduced as positive values.

3. RELATIONS BETWEEN THE POSITIVE AND NEGATIVE FOCAL PLANES.

As in this combination the cylinders likewise affect each other most when their axes coincide, and least when their axes are diametrically opposed, we may here also fix upon the limits of F_1 and $- F_0$ for $\gamma = 0°$ and $\gamma = 90°$, as in the previous theorem.

When $\gamma = 0°$, or $\cos 2\gamma = + 1$, from the equations (48) and (50) we find, for $f_0 > f_1$,

$$F_1 = \frac{f_1 f_0}{\sqrt{\tfrac{1}{2}[-f_1(f_0 - f_1) + (f_0 - f_1)(f_0 + f_0 - f_1)]}} = \frac{f_1 f_0}{f_0 - f_1}$$

$$- F_0 = - \frac{f_1 f_0}{\sqrt{\tfrac{1}{2}[-f_1(f_0 - f_1) + (f_0 - f_1)(f_0 - f_0 + f_1)]}} = - \frac{f_1 f_0}{0} = - \infty.$$

$$\therefore\ F_1 : - F_0 = \frac{f_1 f_0}{f_0 - f_1} : - \infty. \quad . \quad . \quad . \quad . \quad . \quad . \quad (51)$$

For $F_1 = \frac{f_1 f_0}{f_0 - f_1}$, we have as the refraction $\frac{1}{F_1} = \frac{1}{f_1} - \frac{1}{f_0}$; consequently,

11. When the convex cylinder is of greater refraction than the concave, and their axes are coincident, the positive focal plane will coincide with that focal plane which is defined by the difference of the refractions of the cylinders,* whereas the negative focal plane will be at infinity.

---

* Or the sum of their refractions when taken as positive and negative elements.

Placing $\gamma = 0°$, or $\cos 2\gamma = +1$, in the equations (48) and (50), we have, for $f_1 > f_0$,

$$F_1 = \frac{f_1 f_0}{\sqrt{\tfrac{1}{2}[f_1(f_1-f_0)-(f_1-f_0)(f_0+f_1-f_0)]}} = \frac{f_1 f_0}{0} = \infty.$$

$$-F_0 = -\frac{f_1 f_0}{\sqrt{\tfrac{1}{2}[f_1(f_1-f_0)-(f_1-f_0)(f_0-f_1+f_0)]}} = -\frac{f_1 f_0}{f_1-f_0}.$$

$$\therefore\ F_1 : -F_0 = \infty : \frac{f_1 f_0}{f_1-f_0}. \quad \ldots \quad (52)$$

For $-F_0 = -\frac{f_1 f_0}{f_1-f_0}$, we have as the refraction

$$-\frac{1}{F_0} = -\left(\frac{1}{f_0}-\frac{1}{f}\right)\text{; consequently,}$$

12. When the concave cylinder is of greater refraction than the convex, and their axes are coincident, the negative focal plane will coincide with that focal plane which is defined by the difference of the refractions of the cylinders,* whereas the positive focal plane will be at infinity.

This is shown in Plate IV, Fig. 1.

Introducing $\gamma = 90°$, or $\cos 2\gamma = \cos 180° = -1$ in the equations (48) and (50), we have, for $f_0 \gtreqless f_1$,

$$F_1 = \frac{f_1 f_0}{\sqrt{\tfrac{1}{2}[f_1(f_1+f_0)+(f_0-f_1)(f_0+f_0+f_1)]}} = \frac{f_1 f_0}{f_0} = f_1.$$

$$-F_0 = -\frac{f_1 f_0}{\sqrt{\tfrac{1}{2}[f_1(f_1+f_0)+(f_0-f_1)(f_0-f_0-f_1)]}} = -\frac{f_1 f_0}{f_1} = -f_0.$$

$$\therefore\ F_1 : -F_0 = f_1 : -f_0. \quad \ldots \quad (53)$$

*Or the sum of their refractions when taken as positive and negative elements.

From which we deduce :

13. The positive and negative focal planes coincide with their correlative elementary focal planes, when the axes of the contra-generic cylinders are at right angles to each other.

This is demonstrated in Plate IV, Fig. 2.

Between the limits of 0° and 90°, for $f_0 > f_1$, we have consequently found F_1 to vary between the limits of $\frac{f_1 f_0}{f_0 - f_1}$ and f_1 behind the combined lenses, while F_0 varies between the limits of ∞ and f_0 on the incident side of the same.

The convex being stronger than the concave cylinder, it is evident when their axes coincide that their combined refraction will be equal to that of a periscopic convex cylinder, since $\frac{1}{F_1} = \frac{1}{f_1} - \frac{1}{f_0}$ in the active plane; and $\frac{1}{F_0} = \frac{1}{\infty} = 0$ in the passive plane.

Between the same limits, when $f_1 > f_0$, F_0 will vary between $\frac{f_1 f_0}{f_1 - f_0}$ and f_0 on the incident side of the combined cylinders, while F_1 varies between ∞ and f_1 behind the same. (See Plate IV.)

In this case, when the axes coincide, it is evident that the resulant refraction will be equal to that of a periscopic concave cylinder, since $-\frac{1}{F_0} = -\left(\frac{1}{f_0} - \frac{1}{f_1}\right)$ in the active plane; and $\frac{1}{F_1} = \frac{1}{\infty} = 0$ in the axial plane.

Therefore, with an inequality in the refractive powers of the cylinders, rotation of one of them, from 0° to 90°, will be associated with corresponding changes in the position of the resultant focal planes, between the limits of infinity and the focus of the weaker cylinder on the one side, and between that focal plane which corresponds to the difference of their refractions and the focus of the stronger cylinder on the other. Since in this case the approach of one focal plane is accompanied by a corresponding recession of the other on the opposite side of the lenses, their movements are, as in the previous theorem, in opposite directions.

When the cylinders are of equal refractive power, f_1 being equal to f_0, it follows, from the relation (53), that $F_1 = F_0$, so that between the limits of 0° and 90°, F_1 will vary between infinity and f_1 on the positive side, while F_0 varies between infinity and f_0 on the negative or incident side of the combined cylinders.

Consequently, when the axes coincide, $+ F_1 = + \infty$ and $- F_0 = - \infty$. This is evident, since the refractions of equal convex and concave cylinders, under such circumstances, neutralize each other throughout.

By the previous considerations we therefore here also find :

14. The positive and negative focal planes are conjugate planes, subject to variations of the angle between the axes of the contra-generic cylinders.

The diagram, Plate III, has been constructed in accordance with the foregoing provisions.

For practical purposes, it will be found more convenient to use the formulæ in the next chapter.

DIOPTRAL FORMULÆ

FOR COMBINED

CYLINDRICAL LENSES

PLATE II

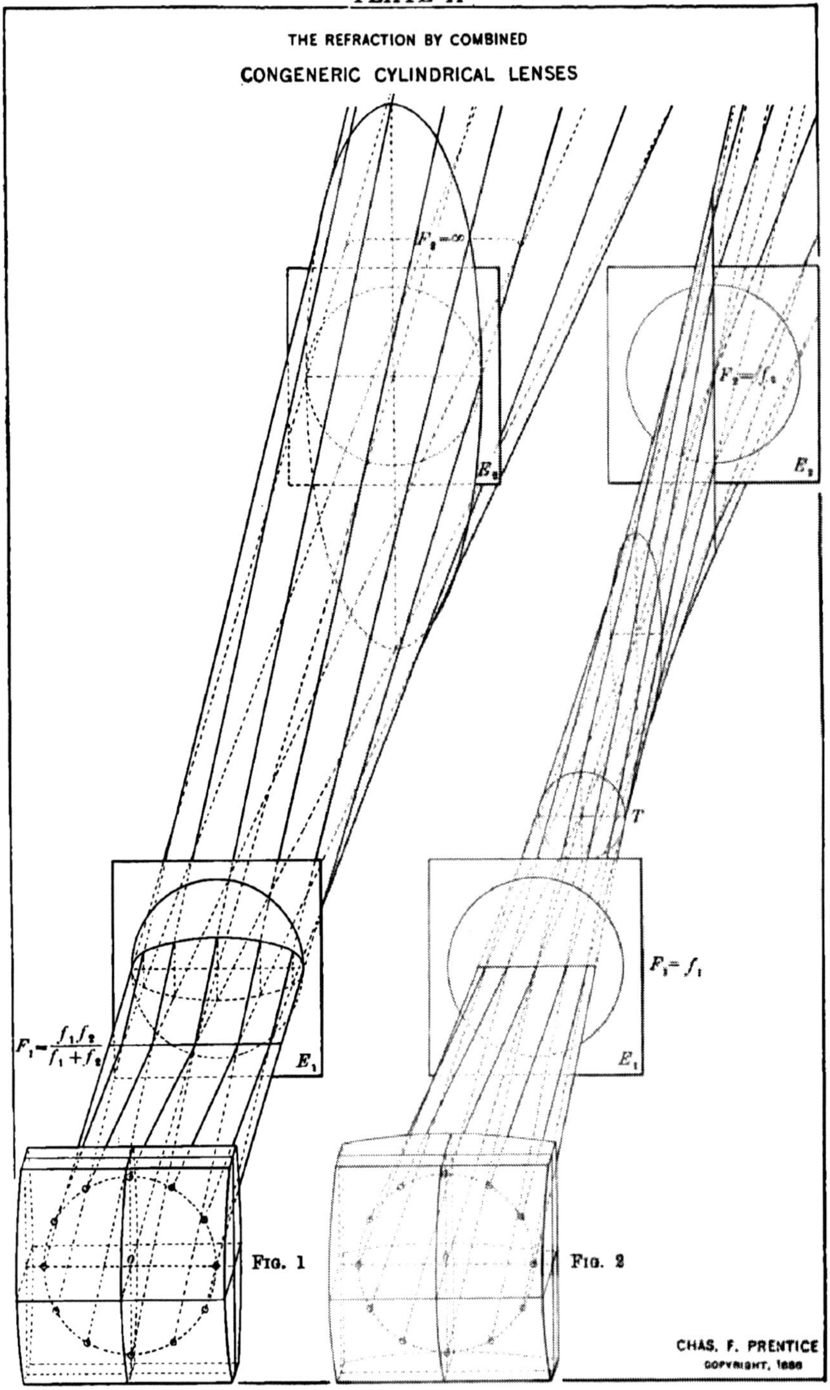

THE REFRACTION BY COMBINED

CONGENERIC CYLINDRICAL LENSES

$F_2 = \infty$

E_2

$F_2 = f_2$

E_2

T

$F_1 = \frac{f_1 f_2}{f_1 + f_2}$

E_1

$F_1 = f_1$

E_1

FIG. 1

FIG. 2

CHAS. F. PRENTICE

COPYRIGHT, 1888

III. DIOPTRAL* FORMULÆ

FOR COMBINED

CYLINDRICAL LENSES.

---

1. RELATION BETWEEN THE PRINCIPAL PLANES OF REFRACTION AND THE REFRACTIVE POWERS OF THE CYLINDERS.

As the task of reducing *dioptries* to their focal distances would render calculation by the preceding formulæ somewhat arduous, we may here give the formulæ, expressed in refraction, which will be found especially convenient when applied to combinations of the metric system.

Since original publication, these formulæ have been given their simplest possible form. The new formulæ, IID, IIID, VIID and VIIID are now introduced as sequences to the original formulæ, which are also given, and whose transformations have been accomplished through convenient substitutions from the equations 54a, 54 and 55.

For the focal distance F_1 we have as the refraction $\frac{1}{F_1} = R_1$, and for f_1 and f_2, similarly, $\frac{1}{f_1} = r_1$ and $\frac{1}{f_2} = r_2$, which designate the dioptral powers of the cylinders.

By these, and similar substitutions for other foci, we may give the preceding formulæ (I) to (X), the following form :

*The choice of this adjective would seem justifiable, since the unit "dioptry" has been chosen in distinction to "dioptric," which, though related, has another significance.

THE DIOPTRAL FORMULÆ FOR COMBINED CONGENERIC CYLINDERS.

$$\cos a = \sqrt{\frac{1}{2} + \frac{1}{2} \frac{r_2 + r_1 \cos 2\gamma}{\sqrt{r_1^2 + 2r_1 r_2 \cos 2\gamma + r_2^2}}}. \quad . \quad . \quad (\mathrm{I}D)$$

$$R_1 = \sqrt{\tfrac{1}{2}(r_1 + r_2)^2 - r_1 r_2 \sin^2\gamma + (r_1 + r_2)\sqrt{\tfrac{1}{4}(r_1 + r_2)^2 - r_1 r_2 \sin^2\gamma}}.$$

$$= \tfrac{1}{2}(r_1 + r_2 + \sqrt{(r_1 + r_2)^2 - 4r_1 r_2 \sin^2\gamma}). \quad . \quad . \quad . \quad . \quad . \quad . \quad (\mathrm{II}D)$$

$$R_2 = \sqrt{\tfrac{1}{2}(r_1 + r_2)^2 - r_1 r_2 \sin^2\gamma - (r_1 + r_2)\sqrt{\tfrac{1}{4}(r_1 + r_2)^2 - r_1 r_2 \sin^2\gamma}}.$$

$$\tfrac{1}{2}(r_1 + r_2 - \sqrt{(r_1 + r_2)^2 - 4r_1 r_2 \sin_2\gamma}). \quad . \quad . \quad . \quad . \quad . \quad (\mathrm{III}D)$$

To retain the significances of R_1 and R_2, in calculating, r_1 should represent the greater cylindrical refraction.

When the cylinders are of equal power, then $r_1 = r_2 = r$, so that

$$R_1 = r(1 + \cos\gamma). \quad . \quad . \quad . \quad . \quad . \quad . \quad . \quad (\mathrm{IV}D)$$

$$R_2 = r(1 - \cos\gamma). \quad . \quad . \quad . \quad . \quad . \quad . \quad . \quad (\mathrm{V}D)$$

THE DIOPTRAL FORMULÆ FOR COMBINED CONTRA-GENERIC CYLINDERS.

$$\cos a = \sqrt{\frac{1}{2} + \frac{1}{2} \frac{r_1 - r_0 \cos 2\gamma}{\sqrt{r_1^2 - 2r_1 r_0 \cos 2\gamma + r_0^2}}}. \quad . \quad . \quad . \quad (\mathrm{VI}D)$$

$$R_1 = \sqrt{\tfrac{1}{2}(r_1 - r_0)^2 + r_1 r_0 \sin^2\gamma + (r_1 - r_0)\sqrt{\tfrac{1}{4}(r_1 - r_0)^2 + r_1 r_0 \sin^2\gamma}}.$$

$$= \tfrac{1}{2}(r_1 - r_0 + \sqrt{(r_1 - r_0)^2 + 4r_1 r_0 \sin_2\gamma}). \quad . \quad . \quad . \quad . \quad . \quad (\mathrm{VII}D$$

$$-R_0 = -\sqrt{\tfrac{1}{2}(r_1 - r_0)^2 + r_1 r_0 \sin^2\gamma + (r_0 - r_1)\sqrt{\tfrac{1}{4}(r_1 - r_0)^2 + r_1 r_0 \sin^2\gamma}}.$$

$$= \tfrac{1}{2}(r_1 - r_0 - \sqrt{(r_1 \quad r_0)^2 + 4r_1 r_0 \sin^2\gamma}). \quad . \quad . \quad . \quad . \quad . \quad (\mathrm{VIII}D)$$

when the cylinders are of equal power, then $r_1 = r_0 = r$, hence

$$R_1 = r \sin \gamma. \qquad \text{(IX}D\text{)}$$
$$-R_0 = -r \sin \gamma. \qquad \text{(X}D\text{)}$$

If, in (IID) and (IIID), the convex element r_2 be replaced by the concave element $-r_0$, we obtain (VIID) and (VIIID).

By the aid of the preceding formulæ we may also arrive at the following significant facts.

The formulæ (IID) and (IIID) may be written :

$$R_1^2 = \tfrac{1}{2}(r_1 + r_2)^2 - r_1 r_2 \sin^2 \gamma + (r_1 + r_2)\sqrt{\tfrac{1}{4}(r_1 + r_2)^2 - r_1 r_2 \sin^2 \gamma},$$
$$R_2^2 = \tfrac{1}{2}(r_1 + r_2)^2 - r_1 r_2 \sin^2 \gamma - (r_1 + r_2)\sqrt{\tfrac{1}{4}(r_1 + r_2)^2 - r_1 r_2 \sin^2 \gamma},$$

which, by addition, result in the equation,

$$R_1^2 + R_2^2 = (r_1 + r_2)^2 - 2r_1 r_2 \sin^2 \gamma.$$

$$\therefore \quad (R_1 + R_2)^2 - 2R_1 R_2 = (r_1 + r_2)^2 - 2r_1 r_2 \sin^2 \gamma.$$
$$\therefore \quad (R_1 + R_2)^2 = (r_1 + r_2)^2 - 2r_1 r_2 \sin^2 \gamma + 2R_1 R_2.$$

Multiplying (IID) by (IIID), we find,

$$2R_1 R_2 = 2r_1 r_2 \sin^2 \gamma. \qquad \text{(54a)}$$
$$\therefore \quad R_1 + R_2 = r_1 + r_2. \qquad \text{(54)}$$

From which we conclude :

15. The sum of the primary and secondary refractions is a constant, being equal to the sum of the elementary refractions for any combination, and all deviations of the axes of two combined congeneric cylinders.

In the same manner, we obtain from the formulæ (VIID) and (VIIID),

$$R_1 - R_0 = r_1 - r_0, \qquad \text{(55)}$$

and therefore here also find,

16. The sum of the principal positive and negative refractions is a constant, being equal to the sum of the positive and negative elementary refractions for any combination, and all deviations of the axes of two combined contra-generic cylinders.

As the total inherent refraction always remains the same for any combination, the angle γ merely performs the function of allotting the proportions of refraction R'_1 and R'_{11}, or R_1 and R_0, in the resultant principal planes.

By the equations (54) and (55), calculation may be greatly simplified. R'_1 being determined for a specific value of γ, we may readily determine R_1 and R'_{11} by transforming these equations, as follows:

$$R'_{1} = r_1 + r_2 - R_1$$

$$- R_0 = r_1 - r_2 - R_1.$$

This is demonstrated in the appended tables, although it has not been utilised in calculating; on the contrary, a study of these led to the above deductions.

SPHERO-CYLINDRICAL EQUIVALENCE

PLATE IV

THE REFRACTION BY COMBINED
CONTRA-GENERIC CYLINDRICAL LENSES
$F_1 = \infty$
$F_1 = f_1$
E_1
E_1
FIG. 1
FIG. 2
$F_0 = f_0$
E_0
E_0
$F_0 = \frac{f_0 f_1}{f_1 - f_0}$
CHAS. F. PRENTICE
COPYRIGHT,

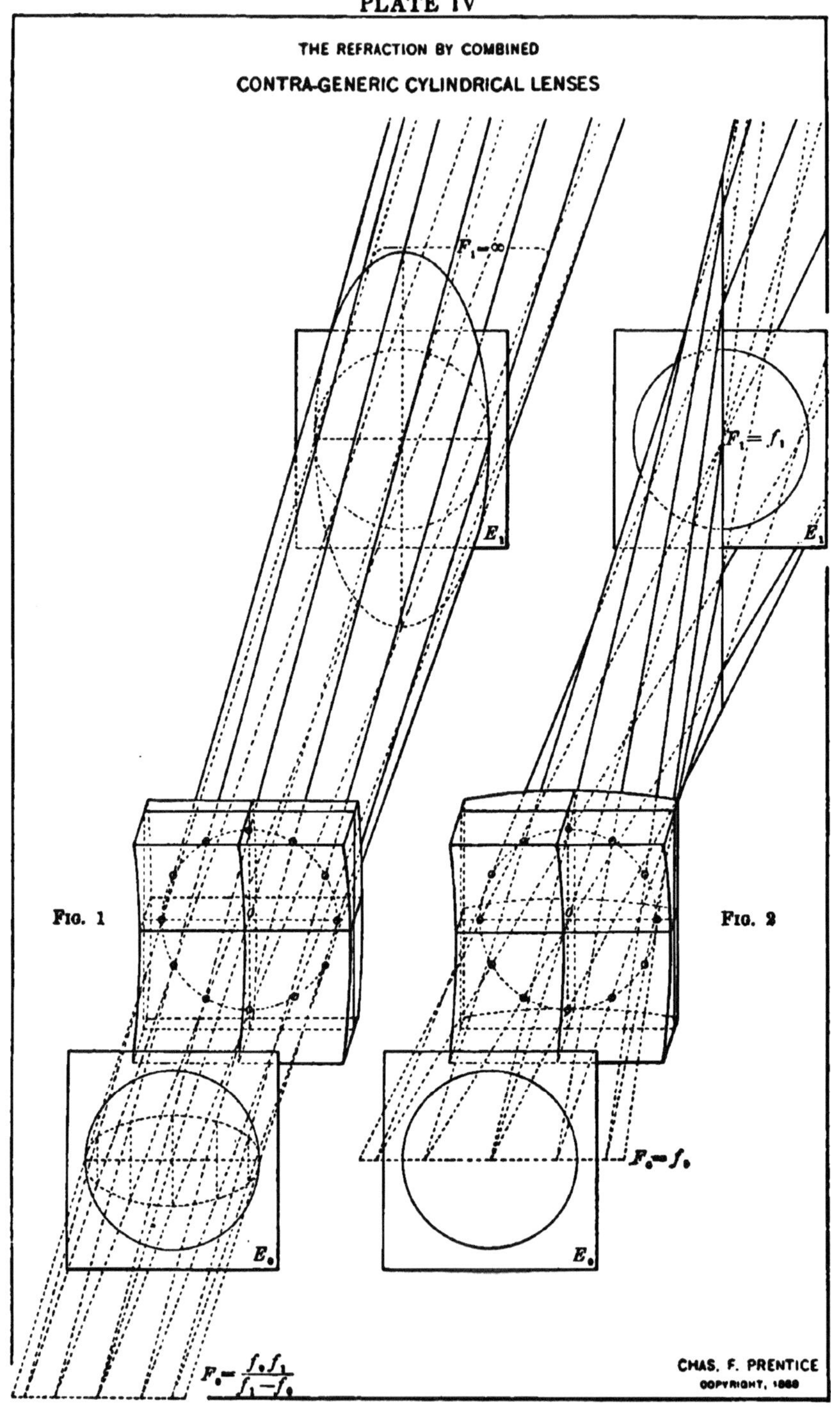

IV. SPHERO-CYLINDRICAL EQUIVALENCE.

Since, for any combination of cylinders, the principal planes of refraction are at right angles to each other for all values of γ, there can be no reasonable doubt, owing to the provisions made at the opening of this demonstration, as to the equivalence of a sphero-cylindrical lens to one composed of combined cylinders. However, as the use of such lenses is at present confined to the correction of errors of refraction in the human eye, it is evident, from the movements of the eye behind the fixed lens, that the visual axis cannot at all times coincide with the optical axis of the lens chosen; therefore, in those instances where substitution of one form of lens for the other proves to be unsatisfactory, the cause might seemingly be explained by a possible difference becoming manifest for the more peripheral incident rays, though these be equally distant from the optical center of each lens. In other words, the available field in the one may be greater or less than in the other; yet even this would probably only be appreciable in lenses of extreme curvature, and possibly in combinations where the cylinders differ widely in power. However, this would remain to be shown.

To substitute a sphero-cylindrical lens for combined cylinders is a proposition which merely demands that the focal interval should be the same, at the same distance from the principal plane, at the optical center, for each of the compound lenses. The distances F_1 and F_2 being determined for any angular deviation γ of the axes, in a combination of congeneric cylinders, for instance, the substitution is accomplished by making a sphero-cylindrical lens in which the focus of the spherical element is equal to F_2, and of the cylindrical element is equal to $\frac{F_1 F_2}{F_2 - F_1}$, or, if expressed by refraction, $\frac{1}{F_1} - \frac{1}{F_2}$ sph. $= \frac{1}{F}$ cyl. $= \frac{1}{F_c}$.

Should it be desired to place the primary and secondary planes of the sphero-cylindrical lens so as to coincide with those resulting from a combination of two definitely placed congeneric cylinders, it will be necessary to refer to the formula (I) and to the laws 2 and 3.

Comparing the sphero-cylindrical equivalent with its corresponding rotating cylinders, reference being had to Plate II, Figs. 2 and 3, we find a decrease in the angle γ from 90° to zero to effect a corresponding decrease in the spherical element F_2, from the focus f_2 to ∞; this being associated with a cylindrical element of the focus F_c, which constantly increases from the focus $\frac{f_1 f_2}{f_2 - f_1}$ to $\frac{f_1 f_2}{f_2 + f_1}$. In other words, a gradually decreasing potency of the spherical refraction $\frac{1}{F_2}$ from $\frac{1}{f_2}$ to $\frac{1}{\infty} = 0$, gives way to a proportionately increasing cylindrical refraction $\frac{1}{F_c}$, from $\frac{1}{f_1} - \frac{1}{f_2}$ to $\frac{1}{f_1} + \frac{1}{f_2}$. As an instance, if $f_1 = f_2 = f$, $\frac{1}{F_c}$ will increase from $\frac{1}{f_1} - \frac{1}{f_2} = 0$ to $\frac{2}{f}$, or twice the refraction of either cylinder. In this case, all successive values of cylindrical refraction will therefore be inherent between 0 and $\frac{2}{f}$.

Should a means be devised to suppress the spherical element for each successive value of γ, the remaining varying cylindrical element being thus rendered available for measuring corresponding degrees of astigmatism in the eye, the formulæ here advanced would prove of service in obtaining the graduations upon the rotating scale of such an instrument.

While there are few cases of astigmation which demand correction by combined cylinders, we may nevertheless be permitted to passingly allude to certain methods of procedure in such instances. We shall confine the subject to congeneric cylinders. In a case of astigmatism which has been found to be corrected by two cylinders combined under the angle γ, the lenses should be withdrawn from the trial frame, when they are to be superposed with their plane surfaces in contact; and in such manner as to facilitate their being rigidly held in the required position for γ.

The positions of the principal planes of refraction may then be estimated for this fixed combination, the same as if it were a single lens, though without regard to the exact nature of the elements constituting it. The powers of the principal planes of refraction will be revealed by neutralizing with the lenses from the trial set. The spherical and cylindrical elements thus determined are then to be substituted in the trial frame, when rotation of the cylinder will lead to that position of it which produces the best acuteness of vision. The spherical and cylindrical elements will probably then also bear of further modification, in case any error may have been made at the outset. In lieu of this practical method, recourse must be had to the formulæ.

It having been shown that successive changes in the angle γ are associated with corresponding changes in F_1 and F_2, the above substitution would indeed seem advisable, since the present appliances for grinding bi-cylindrical lenses are not constructed with sufficient precision to enable opticians to fix the relative positions of the cylinders beyond mere approximation.

As an illustration, let us select two congeneric cylinders of equal foci, say 20 inches, combined under the angle $\gamma = 60°$. Introducing these values in the formulæ (IV) and (V), we find,

$$F_1 = \frac{20}{1 + \cos 60°} = \frac{20}{1 + 0.5} = 13.33,$$

$$F_2 = \frac{20}{1 - \cos 60°} = \frac{20}{1 - 0.5} = 40.$$

We then obtain the cylindrical refraction $\frac{1}{F_c}$, for the desired sphero-cylindrical equivalent, from the equation,

$$\frac{1}{F_1} - \frac{1}{F_2} = \frac{1}{F_c} \quad \ldots \ldots \ldots \quad (56)$$

Substituting herein the calculated values for F_1 and F_2 gives,

$$\frac{1}{13.33} - \frac{1}{40} = \frac{1}{F_c} = \frac{1}{19.99} = \frac{1}{20} \text{ (nearly).}$$

$\frac{1}{F_2} = \frac{1}{40}$ being the spherical element, we therefore have the sphero-cylindrical equivalent,

$$\frac{1}{40} \text{ sph.} \supset \frac{1}{20} \text{ cyl.}$$

as an available substitute for the cylindro-cylindrical lens,

$$\frac{1}{20} \text{ cyl. axis } 0° \frown \frac{1}{20} \text{ cyl. axis } 60°$$

without regard to a definite position of these lenses before the eye.

By way of comparison, allowing the optician to make an error of apparently so small an amount as 2°, in producing the same cylindro-cylindrical lens, we obtain, by introducing $\gamma = 62°$ in the same formulæ,

$$F_1 = \frac{20}{1 + \cos 62°} = \frac{20}{1 + 0.469} = \frac{20}{1.47} = 13.61,$$

$$F_2 = \frac{20}{1 - \cos 62°} = \frac{20}{1 - 0.47} = \frac{20}{0.53} = 37.73.$$

Substituting these values in the equation (56), we have,

$$\frac{1}{13.61} - \frac{1}{37.73} = \frac{1}{F_c} = \frac{1}{21.29},$$

from which we obtain the sphero-cylindrical lens,

$$\frac{1}{37.73} \text{ sph.} \frown \frac{1}{21.29} \text{ cyl.}$$

Had the optician been required to make a sphero-cylindrical lens $\frac{1}{40}$ sph. $\frown$ $\frac{1}{20}$ cyl., his execution of it presenting such discrepancies as $\frac{1}{37.73}$ sph. $\frown$ $\frac{1}{21.29}$ cyl., would certainly be rejected as being unsatisfactory, on account of the notable difference of 2.27 inches in the focal distance of the spherical element.

On the other hand, instances are likely to occur in which it will be impossible, by the advanced method of neutralization to accurately arrive at the sphero-cylindrical equivalent.

Since $\frac{1}{20}$ cyl. axis 0° $\frown$ $\frac{1}{20}$ cyl. axis 62° $= \frac{1}{37.73}$ sph. $\frown$ $\frac{1}{21.43}$ cyl., we should evidently be unable to accurately neutralize such spherical and cylindrical elements by any of the lenses from the trial set.

In those instances, therefore, where satisfactory neutralization of the principal planes of refraction in a pair of combined cylinders cannot be attained, the cylindro-cylindrical lens will have to be chosen, again under the proviso, however, of a faultless mechanical execution. However, as in most instances a sphero-cylindrical equivalent will be available, **we are to suspect error in our estimate of the refraction of an eye which seems to demand cylinders combined under acute or obtuse angles.**

The following is a case in point :

A cylindro-cylindrical lens $-\frac{1}{40}$ cyl. axis 0° $\supset$ $-\frac{1}{40}$ cyl. axis 70° had been prescribed for Mr. G. B. O., of New York, by his oculist in Philadelphia, in 1880–'1. With this lens the vision $= \frac{6}{6}$ for the left eye.

In this instance the sphero-cylindrical equivalent was obtained as follows :

The lenses being congeneric concave cylinders of equal refraction, by the formulæ (IV) and (V), for $f = 40$ and $\gamma = 70°$, we have,

$$F_1 = \frac{40}{1 + 0.34202} = 29.806 = 30,$$

$$F_2 = \frac{40}{1 - 0.34202} = 60.79 = 60,$$

it being admissible to neglect the fractions for such focal distances.

By law 2, we find the position of the cylindrical axis equal $\frac{\gamma}{2} = 35°$, and consequently the sphero-cylindrical equivalent,

$$-\frac{1}{60} \text{ sph. } \supset -\frac{1}{60} \text{ cyl. axis } 35°.$$

This lens was substituted with the knowledge and to the entire satisfaction of the patient.

It is therefore obvious that the meridian (125°) of greatest refraction in the eye had not been disclosed by the oculist's diagnosis.

The weak spherical element in the substituted lens, while being an appreciable factor to the patient, might easily have been overlooked by the practitioner.

In similar cases, the advanced formulæ must prove of value in fixing upon the true state of the refraction.

VERIFICATION

OF THE

DIOPTRIC AND DIOPTRAL FORMULÆ

V. VERIFICATION OF THE FORMULÆ.

In the following tables, the Dioptric and Dioptral Formulæ have been separately applied to combinations of cylinders of the inch and metric systems, respectively. It would be inadmissible to substitute the generally adopted inch-system equivalents for dioptries, in calculating, on account of frequent repetitions of the former as factors in the dioptral formulæ, which would naturally increase the neglected differences to an unwarrantable degree. For the purpose of obtaining reliable results, the calculations have been carried to the fifth decimal point under the radicals. The angles 30°, 45°, and 60° have been chosen so as to exhibit appreciable differences in the corresponding resultant refractions, which are thereby also brought within the lens-series of the inch and metric systems. The elementary foci and refractions have, in a measure, been arbitrarily selected, it being noticeable that the secondary refraction will generally be beyond the limits of neutralization for combinations of weaker cylinders whose axes deviate less than 30°.

The *Approximates* given for refraction, in Table 1, will at times appear to conflict with the laws 15 and 16; this, however, is to be attributed to changes of proportion occasioned by the adopted substitutions.

To substantiate the resultant refractions given in the tables, through the practical test of neutralization, the cylindrical axes should first be accurately determined, and their deviation effectively maintained while the plane surfaces of the cylinders are kept in absolute contact.

In holding these lenses while neutralizing, great care should be exercised to prevent slipping, as the slightest variation in the position of the axes will prove misleading. In this practical experiment, the observer's eye will generally fail to appreciate the neglect of fractions made necessary through using lenses from the trial case.

In explanation of the tables (1) on the following page, under the caption "Elementary Foci," are given, for instance, the foci ($f_1 = 16$ inches, and $f_2 = 24$ inches) of the cylinders whose "Axial Deviation" is 30°. On the same horizontal ruling are given 10.2576 inches as the "Primary Focus," and 149.7422 inches as the "Secondary Focus." The nearest practical equivalents, expressed in refraction, are shown to be $\frac{1}{10}$ and $\frac{1}{160}$, respectively, which in practice will be found to be the lenses most closely *approximating* neutralization of the principal meridians of the combined cylinders.

In the second set of tables (2), under the heading "Elementary Refraction," the cylinders are expressed in dioptries, and in the right hand vertical column the laws mentioned on pages 83 and 84 are forcibly exemplified.

The Dioptral Formulæ on page 82 were applied in these tables, and will generally be found most convenient for use by the student who may desire to solve similar examples. In this event, great care should be exercised to retain the proper meaning and proportions of r_1, r_2 and r_0, as indicated by the respective signs +, —, > and < in the left hand vertical column.

1. TABLES IN VERIFICATION OF THE DIOPTRIC FORMULÆ.

FOR COMBINED CONGENERIC CYLINDERS.

ELEMENTARY FOCI.	AXIAL DEVIATION.	PRIMARY FOCUS.	PRIMARY REFRACTION.	SECONDARY FOCUS.	SECONDARY REFRACTION.
$f_1 < f_2$.	γ	F_1	(Approximate.)	F_2	(Approximate.)
16 ⌒ 24	30°	10.2576	1/10	149.7422	1/160
" "	45°	11.1555	1/11	68 8347	1/72
" "	60°	12.5559	1/12	40.7773	1/40

FOR COMBINED CONTRA-GENERIC CYLINDERS.

ELEMENTARY FOCI.	AXIAL DEVIATION.	POSITIVE FOCUS.	POSITIVE REFRACTION.	NEGATIVE FOCUS.	NEGATIVE REFRACTION.
$f_0 > f_1$.	γ	$+F_1$	(Approximate.)	$-F_0$	(Approximate.)
− 14 ⌒ + 10	30°	16.9799	+ 1/16	32.9799	−1/32
" "	45°	13.2046	+ 1/13	21.2040	−1/22
" "	60°	11.2537	+ 1/11	16.5870	−1/16
$f_0 < f_1$.	γ	$+F_1$	(Approximate.)	$-F_0$	(Approximate.)
− 14 ⌒ + 20	30°	47.5527	+ 1/48	23.5527	−1/24
" "	45°	30.4131	+ 1/30	18.4131	−1/18
" "	60°	23.7316	+ 1/24	15.7315	−1/16

2. TABLES IN VERIFICATION OF THE DIOPTRAL FORMULÆ.

FOR COMBINED CONGENERIC CYLINDERS.

ELEMENTARY REFRACTIONS.	AXIAL DEV'T'N.	PRIMARY REFRACTION.		SECONDARY REFRACTION.		$R_1 + R_2 = r_1 + r_2$
$r_1 > r_2$	γ	R_1	(Approx.)	R_2	(Approx.)	
2.5 ⌒ 1.5D.	30°	3.75D.	3.75D.	0.25D.	0.25D.	4D.
" "	45°	3.46	3.5	0 54	0.5	4
" "	60°	3.09	3.	0.91	1.	4

FOR COMBINED CONTRA-GENERIC CYLINDERS.

ELEMENTARY REFRACTIONS.	AXIAL DEV'T'N.	POSITIVE REFRACTION.		NEGATIVE REFRACTION.		$R_1 - R_0 = r_1 - r_0$
$r_1 > -r_0$	γ	$+R_1$	(Approx.)	$-R_0$	(Approx.)	
+ 4 ⌒ − 2.75D.	30°	2.397D.	+ 2 5D.	1.147D.	−1.25D.	+ 1.25D.
" "	45°	3.052	+ 3.	1.802	−1.75	+ 1.25
" "	60°	3.564	+ 3.5	2.314	−2.25	+ 1.25
$r_1 < -r_0$	γ	$+R_1$	(Approx.)	$-R_0$	(Approx.)	$R_1 - R_0 = r_1 - r_0$
+ 2 ⌒ − 2.75D.	30°	0.856D.	+ 0.75D.	1 606D.	− 1.5D.	− 0.75D.
" "	45°	1.325	+ 1.25	2.075	− 2.	− 0.75
" "	60°	1.690	+ 1.75	2 440	− 2.5	− 0.75

SECTION III

THE PRISM-DIOPTRY

AND OTHER

OPTICAL PAPERS

WITH SIXTY-FIVE ORIGINAL DIAGRAMS

THE PRISM-DIOPTRY.

In the year 1890 the author advocated a "Metric System of Numbering and Measuring Prisms,"* involving the principle that prisms should be numbered according to their refractive powers, instead of by their refracting angles, or angles of minimum deviation. As prisms notably possess the property of apparently changing the position of objects seen through them, it was proposed, in the new system, that the tangent-distance† between the object and its virtual image should form the basis of comparison in measuring the relative strengths of prisms. The tangent-deflection of one centimeter, measured in a plane one meter from the prism, was, therefore, arbitrarily though befittingly chosen as the new unit of prismatic power, and was named the prism-dioptry.

In measuring the refraction of prisms, however, the same as for lenses, it is necessary that the incident pencils of light should be composed of parallel rays, so that the *theoretical* distance of one meter must in practice be increased to at least six meters.

The Prismometric Scale,‡ which is to be placed exactly six meters from the prism, therefore, represents the prism-dioptry as a six-centimeter distance. Scales which are computed for a shorter distance than six meters have been placed upon the market, but, as demonstrated on page 148, are wholly unreliable.

* Archives of Ophthalmology, Vol. XIX, Nos. 1 and 2, 1890; Vol. XX, No. 1, New York, 1891.
Archiv für Augenheilkunde, Band XXII, Berlin, 1890.
The Ophthalmic Review, discussion by Dr. Swan M. Burnett, Vol. X, No. 8, London, 1891.

† Officially adopted by the section of Ophthalmology of the American Medical Association, Washington, D. C., 1891. "To Mr. Prentice alone belongs the credit of having proposed as a standard prism one which produces a deflection of one centimeter at one meter's distance, and no advocate of the centrad ever hinted at it until the appearance of his paper in the *Archives of Ophthalmology*. We owe the simplicity of that idea to Mr. Prentice; let us not deprive him of whatever honor belongs to the conception."—*Medical News*, Philadelphia, May 2, 1891.

‡ The American Journal of Ophthalmology, Vol. VIII, No. 10, St. Louis, 1891.
Les Annales D'Oculistique, Paris, July, 1892.

The author was the first to recommend that the figure of a prism △, used as an exponent* to the prism numerals, should be the symbolic sign for the prism-dioptry, it also being the letter D of the old Greek alphabet. By this means, one prism-dioptry ($1^{\triangle}$) is readily distinguished from the prism of one degree (1°) refracting angle, and, in fact, from prisms of any other system.

The Dioptral† System‡ of numbering prisms alone possesses the great desideratum of establishing a direct and simple relation between the prism-dioptry and the lens-dioptry, as demonstrated by the authors' law,‖ that "a lens decentered one centimeter will produce as many prism-dioptries as the lens has dioptries of refraction." Thus a lens of 1 D. decentered 1 cm. will afford $1^{\triangle}$; a lens of 2 D. decentered 1 cm. will produce $2^{\triangle}$, etc. The

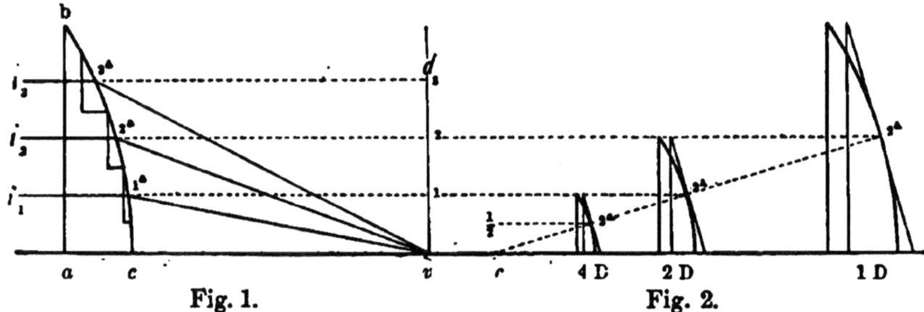

Fig. 1. Fig. 2.

prism-dioptral power is also in direct proportion to the amount of decentration, so that a lens of 2 D. decentered ½ cm. gives $1^{\triangle}$; whereas, if the same lens is decentered 2 cm. it produces $4^{\triangle}$, and so on. It is, therefore, only the size of the lens which in practice will set a limit to its prismatic power.

* Concerning this exponent (△) see paper by Dr. Swan M. Burnett in Annals of Ophthalmology and Otology, July, 1894, Transactions International Ophthalmological Congress, Edinburgh, 1894, and the Refractionist, December, 1894.

The figure of a triangle, no matter how placed in respect to the position of its sides, refers exclusively to the prism-dioptry, being so recognized by American manufacturers.

† First used in Prentice's Dioptric Formulæ for Combined Cylindrical Lenses, monograph, New York, 1888. "The selection of this adjective would seem justifiable, since the unit 'Dioptry' has been chosen in distinction to 'Dioptric,' which, though related, has another significance." Thus, a 40-inch telescope lens is a member of a dioptric system, whereas, a 1-dioptry lens is specifically a member of the dioptral system.

In the English language we have an analogy to dioptry and dioptral in the spelling of ancestry and ancestral.

‡ "Having, by elaborate practical test, fully convinced ourselves of the preëminent advantages of the Dioptral System in the art of manufacture, we have discarded the old degree system entirely, and are now manufacturing prisms which are more accurately ground than ever before." Circular issued to the optical trade by the American Optical Company, Southbridge, Mass.; also catalogue, 1894.

"Our prisms are now ground to conform to the metric system." Catalogue of the Bausch & Lomb Optical Company. Rochester, N. Y., 1895.

‖ Text-book of Ophthalmology, page 141, Drs. Norris and Oliver, Philadelphia, 1893.

Text-book of Diseases of the Eye, page 201, Henry D. Noyes, M.D., New York, 1895.

In Fig. 1, *abc* represents a vertical section of a 1 D. plano-convex lens, with three parallel rays i_1, i_2, i_3, separated by one centimeter distances, which are incident upon its plane side. These rays, after refraction, are collectively directed to the focal point *v*, and therefore suffer perpendicular deflections in the focal plane,* *dv*, which are equal to the correlative decentrations of the rays i_1, i_2, i_3 at their respective points of refraction.

As the spherical surface may be considered as being built up of an unbroken succession of infinitely small prisms of gradually and slightly varying angles, it is to be noted that the three chosen prisms, shown in their order of $1^{\triangle}$, $2^{\triangle}$ and $3^{\triangle}$, correspond to the respective decentrations of 1, 2 and 3 centimeters, and, therefore, produce correlative deflections in the focal plane, *dv*, *exactly* the same as the spherical surface at the same points of refraction. Some recent authors have failed to comprehend this unequivocal precision.

In Fig. 2 three concentric curvatures are shown to represent, respectively, the spherical or cylindrical surfaces of 1, 2 and 4 dioptry lenses, in which the same prism of $2^{\triangle}$ occupies a different position (decentration), relatively to the optical axis, on each of the lenses.

Beneath each section is given the dioptral power of the lens, which, being multiplied by the decentration in centimeters, shows the same prismatic power of $2^{\triangle}$ to exist at a different though definite point on the surface of each lens.

Thus it is seen that every lens, whatever its dioptral power, contains all possible values of the prism-dioptry, which means that the prism-dioptry itself must constitute a distinct part of every lens of the dioptral system.

The prism-dioptry, therefore, stands unchallenged in its unique ability to harmonize all of the refracting elements in the optometrical lens-case by establishing a complete and inseparable relationship between prisms and lenses. We need only to remember the centimeter in connection with the prism-dioptry, as we do the meter in its relation to the lens-dioptry.

* "The great and enduring work of Gauss on the elucidation and simplification of optical laws has among its cardinal elements four planes—the anterior and posterior focal planes and the two principal planes (Haupt-Ebenen); and the proportion of the size of image to object, as elucidated by the formula of Helmholtz, is calculated on the tangent plane. . . . This plane can, in the case of prism-deflection, be regarded in the same light as the focal plane of the standard lens. . . . This method was first suggested and made practical by Mr. C. F. Prentice, of New York, . . . who has gone very thoroughly into the mathematics of the subject in his paper."—*The Ophthalmic Review*, a monthly record of ophthalmic science, London, England, January, 1891.

A METRIC SYSTEM OF NUMBERING AND MEASURING PRISMS.

Revised reprint from the Archives of Ophthalmology, Vol. XIX, No. 1, 1890. Also translated by the author, and published in Archiv für Augenheilkunde, Band XXII, Berlin, 1890.

Introductory Remarks by Dr. Swan M. Burnett.

"The old method of numbering prisms simply by the angular deviation of their sides is, confessedly, inaccurate and unscientific. Any attempt to supplant this by one more accurate, and to place the nomenclature of prisms on the same basis of scientific exactness as the other optical appliances in the hands of the practical ophthalmologist is, therefore, deserving of consideration. The method proposed by Mr. Prentice, in the following paper, not only does this, but does it in a manner and according to principles which are familiar to even the less sicentific practitioners. To have the same unit (the meter) of measure and comparison for all refracting apparatus and uniform with the nomenclature employed in the designation of anomalies of refraction and muscular equilibrium, gives a simplicity which is not only commendable in itself, but tends to render the study of the practical use of prisms easier and more comprehensible to the student. This is particularly apparent in the connection the author establishes between the prism-dioptry, the lens-dioptry and the meter-angle. Not the least important part of the contribution is the description of the instrument Mr. Prentice has devised for illustrating his idea and for testing the refraction of prisms generally."

The present method of designating prisms by the angular deviation of their refracting surfaces, is open to the objection that we thereby define only an isolated feature of their construction, to the utter disregard of the varying powers of refraction, which must result from the use of refracting substances having different indices of refraction.

With a view to securing greater accuracy and uniformity in our utilization of the refractive properties of prisms, the following system of numbering, which the author believes to be feasible, as well as suited to the requirements of optometrical practice, is presented.

Let *abc*, Fig. 1, represent a prism, with the ray *i* incident perpendicularly to *ab*, and we shall have *dv* as the deflection accruing from the refraction at *e*. Similarly, *dV* will represent the deflection arising from the refraction at the same point *e*, for a prism, *ABC*, of greater angle.

We shall then have

$$dV : de = d_1v_1 \ ; \ d_1e$$

But $$d_1v_1 = dv = \delta$$

$$\therefore \frac{dV}{de} = \frac{\delta}{d_1e}$$

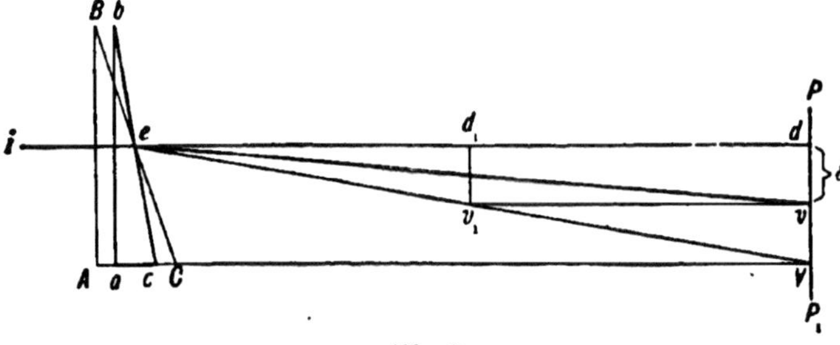

Fig. 1.

By adopting one meter as the distance of the point e from the plane PP_1, in which the amplitudes of deflection dv and dV are measured, we shall have $de = 1$, when

$$dV = \frac{\delta}{d_1e} \quad . \quad . \quad . \quad . \quad . \quad . \quad . \quad . \quad . \quad . \quad (1)$$

It is further evident that the greater the refractive power of the prism, the greater will be its correlative deflection in this plane, so that

$$R : dV = r : \delta \quad . \quad . \quad . \quad . \quad . \quad . \quad . \quad . \quad . \quad (2)$$

where R and r represent the refractive powers of the prisms ABC and abc, respectively.

Consequently, $R = dV$, when $r = \delta$.

Substituting these values in the equation (1), we have

$$R = \frac{1}{d_1e} \cdot r$$

or, if $d_1e = x_m$ as a function of the meter,

$$R = \frac{1}{x_m} \cdot r$$

Allowing r to represent the unit of prismatic refraction, we then have

$$R = \frac{1}{x_m}$$

Consequently, the prismatic refraction is in inverse proportion to the distance at which the unit deflection is produced, being fully in harmony with the refraction of a lens, which is in the inverse proportion to the distance at which the image is formed.

Provided, therefore, a standard amplitude of deflection be adopted as the unit, and which shall be measured in a plane one meter from the refracting surface of the prism, we shall be enabled to designate prisms, in specific terms of dioptries, for instance, with the same significance as in lenses.

Thus prisms of, say, two, three or four prism-dioptries will produce the same unit-deflection at one-half, one-third, or one-quarter of a meter, respectively.

By the relation (2) we shall also find the same prisms to produce two, three or four times the unit-deflection *in the meter-plane*, so that the problem reduces itself to the selection of a series of prisms which shall produce tangent-deflections, at this distance, which are multiples of the adopted unit.

As the deviation produced by a prism will be dependent upon its angle and the index of refraction, we may here note the relation existing between these factors, when, as in Fig. 2, a ray i is incident perpendicularly to the face ab, and, consequently, suffering refraction at e only. Here we

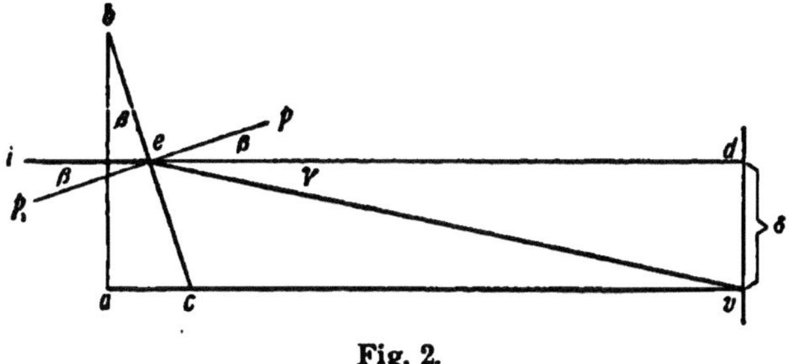

Fig. 2.

have $\sphericalangle abc = \beta$, as the angle of the prism; $\sphericalangle iep_1 = \beta$, the angle of incidence; $\sphericalangle pev$, the angle of refraction; $\sphericalangle dev = \gamma$, the angle of deviation.

Consequently the index of refraction

$$\eta = \frac{\sin \sphericalangle pev}{\sin \sphericalangle iep_1} = \frac{\sin (\sphericalangle ped + \sphericalangle dev)}{\sin \sphericalangle iep_1} = \frac{\sin (\beta + \gamma)}{\sin \beta}$$

As the angles of prisms which we shall here have to consider are comparatively small, we shall use the above instead of the formula for the minimum deviation, which is

$$\eta = \frac{sin \frac{(\beta + \gamma)}{2}}{sin \frac{\beta}{2}}.$$

A relation, therefore, exists between η, β and γ, which requires that two of these factors be known to enable us to determine the third, so that our choice of a unit deflection will be included in the following propositions, wherein prisms of low degree, the usual limits of refractive index, and comparatively small deviations of the refracted ray will have to be considered:

Proposition I.—The values of η and β being given to find γ
" II.— " " " η " γ " " " " β
" III.— " " γ " β " " " " η

For purposes of illustration and reference, we may consider only the first proposition, for the generally accepted index, $\eta = 1.53$ (Spiegelglass), for a 1° prism, when we have

$$\frac{sin (1° + \gamma)}{sin 1°} = 1.53 \therefore sin (1° + \gamma) = 1.53 \times 0.017452 = 0.026701 = sin 1° 31'.8$$

$$sin (1° + \gamma) = sin 1° 31' 48''$$

$$\therefore \gamma = 31' 48''$$

The deflection δ, corresponding to the angle γ, being equal to

$$de. tang \gamma = 1_m tang \gamma = tang 31' 48''$$

we have, in meters, $\delta = 0.009250.$

A prism producing a deflection equal to the tangent of 31′ 48″, equal to 0.00925 at a distance of one meter, will therefore correspond to the accurately ground prism of one degree refracting angle, with an index of 1.53.

In case glass of another index were used, it would be necessary to vary the angle of the prism, so as to satisfy the conditions of refraction for producing the aforesaid deflection, and it is therefore obvious that manufacturers will be privileged to adopt any correlative proportion of angle to index which will satisfy the demands for any tangent-deflection which it may be determined to adopt as a unit.

Supposing the chosen unit of deflection, tang γ, to be slightly greater than the above, say exactly equal to 0.01, or one centimeter, our series of prisms would then be :

1	Prism-dioptry	producing	a tangent	deflection	=	1^{cm}	in the	meter-plane.
2	"	"	"	"	=	2^{cm}	"	"
3	"	"	"	"	=	3^{cm}	"	"

A system of numbering prisms in terms of prism-dioptries, could therefore be adopted which would satisfy all the conditions here set forth.

Such prisms could be measured by noting the deflection they produce upon the index line of a coarse centimeter scale, placed at right angles to the line of sight, at the distance of one meter.* (See Fig. 3.)

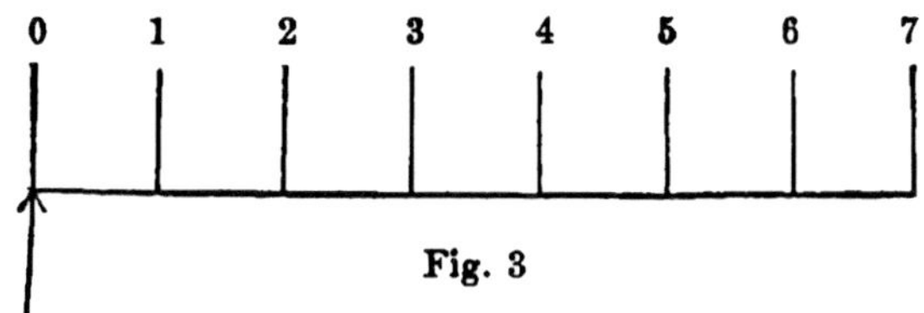

Fig. 3

While a restriction of this character offers the advantage of a ready ocular means of verifying the correctness of the prisms, there are at present however, many difficulties to be overcome in manufacturing them. Calculation would disclose the fact that such prisms would require to be ground to degrees, minutes and seconds, so that comparatively few prisms out of a lot, at the close of our effort to produce them, would be found to actually meet the aforesaid requirements.

This would so heighten their cost as to render them impracticable, except as diagnostic instruments in the consultation room.

Even these, however, could be substituted by the prism-mobile, which consists of two prisms rotating before each other in opposite directions, and

* For reasons given in the paper on the Prismometric Scale, perfect accuracy will only be insured when this scale is enlarged to the dimensions required for its use at a distance of six meters.

which will afford the most ready means of filling a demand for definite deflections, inasmuch as the rotation of the prisms, from 0° to 180°, produces all possible deflections from one millimeter upward.

The instrument could easily be graduated to read to centimeters, and tenths, or millimeters of deflection. For the determination of muscular insufficiences of comparatively low degree, and to render the instrument as light as possible, a special cell, to contain weak rotating prisms, could be devised, similar to that of Dr. Risley, to fit the trial frame.

We may venture to assert that the prism, although the simplest element in Dioptrics, is the most difficult to manufacture, when required to be *exact;* and we shall therefore be obliged, for the present at least, to use existing commercial prisms for spectacle glasses.

We shall subsequently show that these may be profitably utilized, by assigning the unavoidable variations of deflection, consequent to the manufacture of such prisms, to their proper places, as members of the new system.

By actual experiment the author has found imported prisms, represented as being of one degree (1°), to produce deflections varying between 9 and 12 millimeters, and which, if reduced to the basis of our standard of one centimeter, are to be designated as 0.9 and 1.2 prism-dioptries, respectively.

Similar discrepancies of deflection are found to exist throughout the entire series of imported prisms now in use, so that we shall have adequate variety, covering almost every required interval of the new system; whereas, by the earlier method, an optical prescription, although required to be exact, is constantly exposed to the danger of being reduced to little else than a ticket of chance in an optical lottery.

Farther on an instrument will be described which the author has devised to determine the power of prisms in dioptries and fractions, thus making it possible for the optician to select from his stock the one that shall fill the requirements of deflection sought. Indeed, by its use, we may hope to have manufacturers ultimately furnish us prisms, in packages, assorted and marked with the number indicating their power in dioptries.*

* This expectation has been fulfilled by American manufacturers ever since 1891, see page 102.

THE RELATION OF THE PRISM-DIOPTRY TO THE METER-ANGLE.

In the accompanying Fig. 4, E_1 and E_2 represent the centers of rotation of the eyes, and OM the median line, bisecting the base-line $E_1 E_2$ at M.

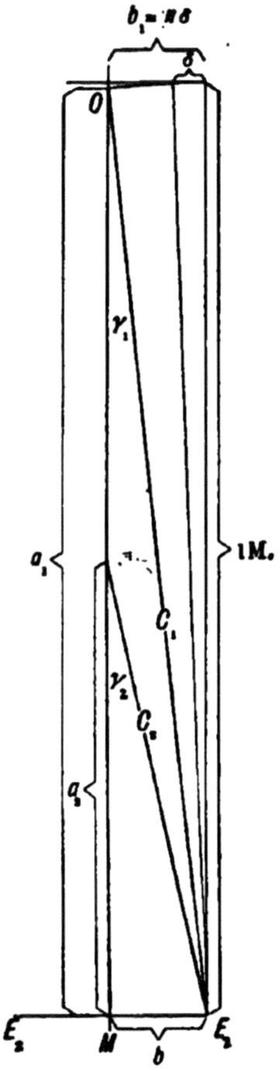

For the point of fixation O, corresponding to the angle of convergence, γ_1 for the eye E_1, we have

$$sin\ \gamma_1 = \frac{b}{OE_1} = \frac{b}{C_1}$$

or if $C_1 = 1$ meter, $sin\ \gamma_1 = b =$ the deviation required of the eye which is optically adapted for the point O.

The base-line, b, having a constant value for each inter-pupillary distance, $2b$, we have the angle γ_1 solely dependent upon the varying metrical values of C_1, so that the unit-angle of convergence has been designated, by Nagel, the meter-angle, when $C_1 = 1$ meter. Hence $1\ ma = arc\ sin \frac{b}{C_1} = arc\,sin \frac{b}{1} = b$;-
$2\ ma = arc\ sin \frac{b}{C_2} = arc\ sin\ \frac{b}{\frac{1}{2} C_1} = arc\ sin \frac{b}{} = 2\,b$;-
a. s. f.

We may now proceed to find the value of the prism in prism-dioptries, which, being placed before the eye, with its base in, shall substitute an effort of convergence to the same point.

The deviation produced by a prism of one prism-dioptry, at one meter, being equal to δ, Fig. 4, it is evident that the prism which shall be required to produce a deviation $b = b_1 = \eta\delta$, will have to be n times greater than one prism-dioptry (1^{Δ}),† consequently $= \eta . 1^{\Delta}$.

Fig. 4.

If the same deviation is to be produced at the distance $a_1 = \frac{1}{Y}$ of one meter, when $C_1 = 1$ meter, for the meter-angle, the prism will require to be Y times as great, consequently

† See page 102.

$$1\ ma = Y.\eta.1^{\Delta}$$

$$\text{But } Y = \frac{1}{a_1} \text{ and } \eta = \frac{b}{\delta}$$

$$\therefore 1\ ma = \frac{1}{a_1} \cdot \frac{b}{\delta} \cdot 1^{\Delta}$$

$$\therefore 1\ ma = \frac{tang\ \gamma_1}{\delta}.\ 1^{\Delta}$$

Similarly, if the deviation b is to be produced at the distance $a_2 = \frac{1}{Z}$ of one meter, when $C_2 = \frac{1}{2}$ meter, for two meter angles, the prism will require to be Z times greater than $\eta.1^{\Delta}$, or $Z.\eta.1^{\Delta}$. Consequently

$$2\ ma = Z.\eta.1^{\Delta}$$

$$\text{But } Z = \frac{1}{a_2} \text{ and } n = \frac{b}{\delta}$$

$$\therefore 2\ ma = \frac{1}{a_2} \cdot \frac{b}{\delta} \cdot 1^{\Delta}$$

$$\therefore 2\ ma = \frac{tang\ \gamma_2}{\delta}.\ 1^{\Delta}$$

$$\therefore Xma = \frac{1}{a_x} \cdot \frac{b}{\delta} \cdot 1^{\Delta} = \frac{tang\ \gamma_x}{\delta}\ 1^{\Delta} \quad . \quad . \quad . \quad \text{(I)}$$

When convergence is confined to comparatively small angles, we may regard the sine, tangent, and angle as being equal to one another. In other words, we may consider $a_1 = C_1 = 1$, $a_2 = C_2 = \frac{1}{2}$, etc., so that

$$1\ ma = \frac{1}{a_1} \cdot \frac{b}{\delta} \cdot 1^{\Delta} = \frac{b}{\delta} \cdot 1^{\Delta}$$

$$2\ ma = \frac{1}{a_2} \cdot \frac{b}{\delta} \cdot 1^{\Delta} = 2\ \frac{b}{\delta} \cdot 1^{\Delta}$$

$$Xma = \frac{1}{a_x} \cdot \frac{b}{\delta} \cdot 1^{\Delta} = X\ \frac{b}{\delta} \cdot 1^{\Delta} \quad . \quad . \quad \text{(II)}$$

Under these circumstances one prism-dioptry differs from the meter-angle by the co-efficient $\frac{b}{\delta}$. This is due to the selection of a comparatively small unit-deflection δ. If we had chosen a greater unit-deflection, say $\delta = b$, then one meter-angle would correspond to one prism-dioptry exactly. A prism, producing so great a deflection as half the inter-pupillary distance, would, however, give too great an angle for the unit and lowest degree of prism, unless others, as fractions of the unit, were included in the series. For instance, if $\delta = b$ equal the deflection for one prism-dioptry, prisms of lesser refraction might be designated as 0.25^{Δ}, 0.5^{Δ}, 0.75^{Δ}, when their respective deflections are $\frac{1}{4}\ b$, $\frac{1}{2}\ b$, and $\frac{3}{4}\ b$.

Such a selection would, however, possess no particular advantages, since b will generally be a variable quantity for different individuals; besides, it will not be admissible to approximate the factor $\frac{1}{a_x}$ for considerable degrees of convergence.

A strict consideration of the co-efficient $\frac{b}{a_x\ \delta}$, under such circumstances, will be imperative, without regard to any particular choice of the unit deflection, so that for considerable degrees of convergence the prism-dioptries will have to be determined by the Formula I.

The following table exhibits the errors committed in estimating the value of the angle of convergence, when Formula II is substituted for Formula I, in reducing meter-angles to prism-dioptries.

METER-ANGLES REDUCED TO PRISM-DIOPTRIES, FOR AN INTER-PUPILLARY DISTANCE OF 64 MILLIMETERS ($b = 32$ m/m). STANDARD UNIT DEFLECTION FOR 1 PRISM-DIOPTRY $= \delta = 0.01 = 1^{cm}$ AT THE METER-PLANE.

Distance from the Object of Fixation to the Eye.		Sine of the Meter-Angle.	Tangent of the Meter-Angle.	Value of the Angle of Convergence.			Value of the Angle of Convergence when sine, tangent and angle are accepted as equal.			
In Meters.	In Millimeters.			In Meter-Angles.	In Prism-Dioptries.	In Degrees exactly = Δ Deviation.	In Meter-Angles.	In Degrees arc sin.	In Prism-Dioptries.	Δ Deviation arc tang.
1	1,000	0.032	0.0320163	1	3 20163	1° 50′ 1″5	1	1° 50′ 1″5	3.2	1° 49′58″2
$\frac{1}{2}$	500	0.064	0.0641309	2	6.41309	3° 40′ 9″9	2	3° 40′ 3″	6.4	3° 39′48″
$\frac{1}{3}$	333.3	0.096	0.0964465	3	9.64465	5° 30′ 32″ 1	3	5° 30′ 4″5	9 6	5° 29′ 0″9
$\frac{1}{4}$	250	0.128	0.129061	4	12 9061	7° 21′ 14″ 4	4	7° 20′ 6″	12.8	7° 17′38″9
$\frac{1}{5}$	200	0.160	0.162080	5	16.2080	9° 12′ 24″ 6	5	9° 10′ 7″5	16	9° 5′24″8
$\frac{1}{10}$	100	0.32	0.337755	10	33.7755	18° 39′ 46″ 7	10	18° 20′15″	32	17° 44′40″
$\frac{1}{20}$	50	0.64	0.832919	20	83.2919	39° 47′ 30″	20	36° 40′30″	64	32° 37′ 8″5

It is apparent that such a substitution will be admissible, up to five meter-angles, where the difference between the deviation produced by the prism, and the value in degrees of five meter-angles amounts only to 4′ 42″ 7.

When we consider that a muscular insufficiency of five meter-angles is entirely beyond the limits of optical correction, the latter, in fact, being confined to deficiencies of about one meter-angle, or less, it is obvious that a substitution of the sine for the tangent will be justifiable.

Such being the case, the subject of prismatic corrections becomes wonderfully simple. For instance, for an inter-pupillary distance of 60 millimeters $= 6^{cm}$, the base line will be 3^{cm}, when, according to Formula II,

$$1\ ma = \frac{3\,cm}{1\,cm} = 3 \text{ prism-dioptries.}$$

Similarly, for an inter-pupillary distance of 50 millimeters, the base-line being 25 millimeters, equal to 2.5 centimeters, the meter-angle will be equal to 2.5^{Δ}.

Thus, for each inter-pupillary distance, we find a different prism necessary to supplant the meter-angle. This is but natural, since greater demands for convergence will be necessary in wide than in narrow inter-pupillary distances.

This leads us to the final and simple rule :

Read the patient's inter-pupillary distance in centimeters, when half of it will indicate the prism-dioptries required to substitute one meter-angle for each eye.

One could scarcely hope for a more convenient method than to find the prism-dioptries, corresponding to one meter-angle, expressed in the patient's features.

There will, however, frequently be occasion to supply less than one meter-angle, as indicated in the following tabulated examples.

Pupillary distance, $2b$ =	56	60	64	68	millimeters
Base-line . . . b =	2.8	3	3.2	3.4	centimeters
1 meter-angle . . . =	2.8	3	3.2	3.4	prism-dioptries
$\frac{1}{3}$ meter-angle . . . =	0.93	1	1.06	1.13	" "
$\frac{1}{2}$ meter-angle . . . =	1.4	1 5	1.6	1.7	" "

According to our standard, a prism of 0.9^{Δ} will produce a tangent deflection of 0.9 of a centimeter, or 9 millimeters, and a prism of 1.1^{Δ} a deflection equal to 1.1 centimeters, or 11 millimeters. It will therefore be possible to select these, by aid of an adequate instrument, from a paper of 1° prisms of foreign manufacture, since the latter are found to produce deflections varying between the same limits.

Prisms of 1.3^{Δ} to 1.6^{Δ} will similarly be found among prisms of $1\frac{1}{2}°$, and so on.

Later we shall describe the instrument, called a prismometer, to distinguish it from optical theodolites and goniometers of the physical laboratory, which, we believe, offers an advantage over the present methods of measuring prisms, inasmuch as it makes it possible to measure prisms to the nicety of fractions.*

The general principle evolved, therefore, affords a new means of verifying the correctness of prisms in a simple manner, and must assuredly serve its purpose, whichever standard unit-deflection, at one meter distance, it may be officially determined to adopt, although it is believed that the centimeter has been shown to possess such decided advantages as to be worthy of favorable consideration.

THE RELATION OF THE PRISM-DIOPTRY TO THE LENS-DIOPTRY.

Some of the advantages of the prism-dioptry, as the unit of measure for the refraction of simple prisms, having been shown, and whereas prisms are frequently combined with spherical lenses, it is here proposed to further consider the relations of the prism-dioptry to such combinations, as well as to the equivalents which are to be obtained by a mere decentration of the spherical lenses themselves.

In the accompanying figure 5, a lens is shown with its principal anterior and posterior foci *f* and *F*, equidistant from *O*, upon the optical axis *fOF*.

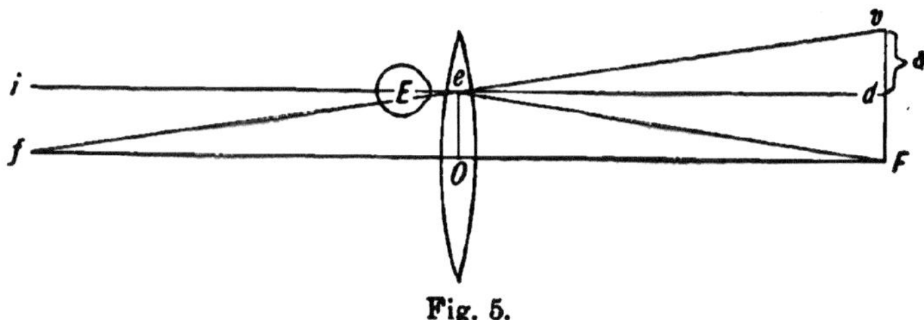

Fig. 5.

The ray, *ie*, which is parallel to the optical axis, and incident at an eccentric point *e*, of the lens, being refracted to the focal point *F* will sus-

* In the original paper, the first, and consequently more or less crude prismometer, was described as being capable of measuring prisms up to 20△. As in subsequent papers this was shown to have been an error, the reader is referred to the descriptions of the Perfected Prismometer, and the Prismometric Scale.

tain a deflection dF, *in the focal plane*, which will always be equal to the decentration Oe.

A ray, de, which is parallel to the optical axis and incident from the opposite direction, will be refracted to the focal point f, and, if received by the eye at E, will be projected by it in the prolongation of fe to v.

Consequently $dv = \delta$ will be the measure of the apparent displacement of the point d, resulting from the prismatic action inherent at e. By reference to the figure, and previous definitions, we then have

$$\sphericalangle\ dev = \sphericalangle\ Ofe$$
$$\sphericalangle\ deF = \sphericalangle\ OFe$$
$$\text{But } \sphericalangle\ Ofe = \sphericalangle\ OFe$$
$$\therefore\ \sphericalangle\ dev = \sphericalangle\ deF$$
$$\therefore\ \delta = dF = Oe$$

The tangent deflection δ, *at the focal plane of the lens*, is therefore always equal to the amount of decentration, and consequently in direct proportion to it.

Augmented decentration of the lens will be associated with an increase in the prismatic action, resulting from a growing inclination of the tangents t_1 t_2 . . . determining the obliquity of the spherical surfaces, at corresponding opposite points of eccentricity, as shown in Fig. 6.

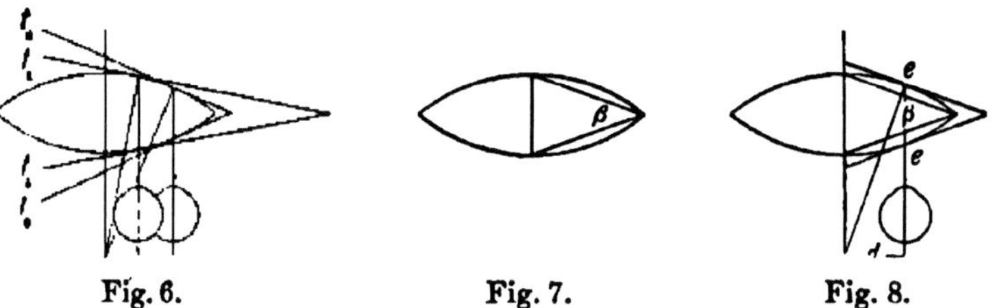

Fig. 6. Fig. 7. Fig. 8.

The inclination of these tangents, relatively to each other, becomes a maximum when the angle between them reaches 180°, that is to say, when they both coincide, and so form the tangent to the lens, which must then have become a perfect sphere. The amount of prismatic action it will be possible to obtain consequently only depends upon the diameter of the lens. It will be well, however, to bear in mind that lenses of large diameter, whose spherical aberration for peripheral rays causes the latter to fall short of the focus, will naturally also effect a still further increase in the prismatic action at the focus, for an extreme decentration.

It is further obvious that a *virtual* prism of *constant* angle cannot exist within the lens to produce the aforesaid variable results. The Fig. 7 is therefore apt to prove misleading, as there is but one fixed amount of decentration, *d*, Fig. 8, which corresponds to a prism of the angle β, and which is determined by the tangents to the lens surfaces at *e*, drawn parallel to the sides of the inscribed prism.

Thus prepared we may proceed to a consideration of the prism-dioptry.

Supposing a lens to be 1 D., and the deflection sought to be $1^{cm} = 1^{\Delta}$ at the meter-plane, which is the focal plane of the lens, it would require a decentration of one centimeter to produce this result.

A lens of 2 D., being decentered the same amount, would produce a tangent deflection equal to 1^{cm} *at its focus, half a meter*, which would be equivalent to 2^{cm} at the meter-plane, or 2^{Δ}, it having been shown that the prismatic refraction is in the inverse proportion to the distance at which the unit-deflection is produced. The following tabulation will serve to make this clear.

Lens.	Decentration.	Tang. Deflections at the Focus.	Tang. Deflections at the Meter Plane.
1 D.	1^{cm}	1^{cm} at 1 meter	$1^{cm} = 1\Delta$
2 D.	1^{cm}	1^{cm} at $\frac{1}{2}$ meter	$2^{cm} = 2\Delta$
3 D.	1^{cm}	1^{cm} at $\frac{1}{3}$ meter	$3^{cm} = 3\Delta$

This table also reveals the unique law that:

Any lens is capable of producing as many prism-dioptries as the lens possesses dioptries of refraction, provided it is decentered one centimeter.

On the other hand the prism-dioptries will increase or decrease as the decentration becomes greater or less. Thus :

Lens.	Decentration in Centimeters.			Decentration in Millimeters.		
	1 cm.	2 cm.		1 m/m	3 m/m	6 m/m
0.25 D.	0.25	0.5	—Prism-Dioptries—	0.025	0.075	0.15
0.5 D.	0 5	1	" "	0.05	0.15	0.3
0.75 D.	0.75	1.5	" "	0.075	0.225	0.45
1 D.	1	2	" "	0.1	0.3	0..6
2 D.	2	4	" "	0.2	0.6	1.2

A lens of 2 D., limited by its size to a decentration of $3^{m}/_{m}$, will afford 0.6^{Δ}, whereas a lens of 1 D., capable of a decentration of $6^{m}/_{m}$, will produce the same prismatic effect, as shown above. In other words, a lens of one-half or one-third the power will require to be decentered twice or three times as much to secure the same number of prism-dioptries.

This we find graphically demonstrated in the accompanying figures, in which the dimensions of decentration and lens-curvatures are exaggerated to better serve our purpose.

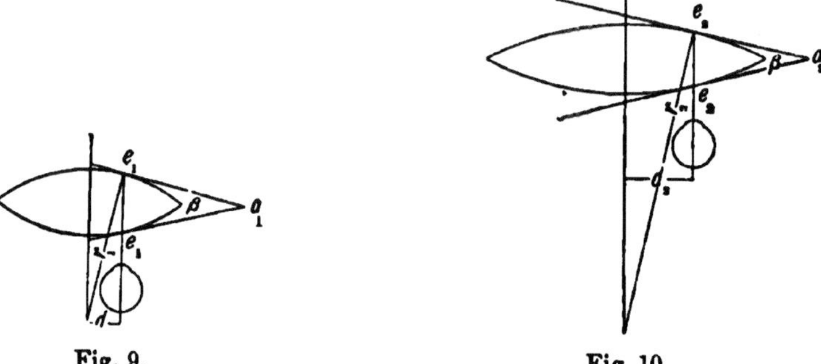

Fig. 9. Fig 10.

In Fig. 9 the lens of 2 D., with a decentration $d_1 = 3\ ^{m}/_{m}$, produces a deflection of 0.6^{Δ} *at the meter plane*, by reason of the obliquity of the lens-surfaces, determined by the tangents at e_1 constituting a *virtual* prism of the angle β, with its apex at a_1.

In Fig. 10 the lens of 1 D., whose radius $r_2 = 2\ r_1$, is decentered by the amount $d_2 = 2\ d_1 = 6^{m}/_{m}$, to produce 0.6^{Δ} *at the meter plane* by a virtual prism of exactly the same angle β, with its base at $e_2\ e_2$, and apex at a_2.

We have consequently but to remember that:

The prism-dioptries in decentered lenses are in direct proportion to their refraction and decentration.

It is therefore actually possible to determine the dioptral power of any pair of contra-generic lenses which neutralize each other, and whose power may be unknown. All that is necessary is to place the lenses over each other, and to separate their optical centers *exactly* one centimeter, when the prism-dioptral power, as read from the Prismometric Scale*, will be equal to the dioptral power of the lenses.

* See method of manipulation on page 140.

Our present lenses will, however, not permit of a decentration of 1cm, owing to their limited size, yet, if required, larger ones could be furnished to the trade, at a comparatively small increase in cost, which could be utilized specifically in corrections involving prism-dioptries.

However, in cases where inadequate size of the lens prevents decentration, it becomes necessary to add to it a constant prism. Since the slightest decentration of a simple lens is certain to produce a prismatic effect, it is evident that the *constant* value of any prism, upon one of whose surfaces a spherical lens has been ground, will only be retained when the optical axis of the lens strictly coincides with the visual axis. See Fig. 11.

The effect of decentration will naturally be to increase or diminish the prismatic action of the constant prism, which has been combined with the lens.

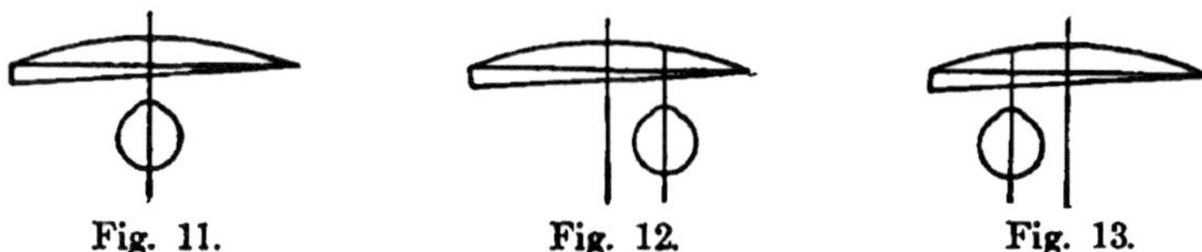

Fig. 11. Fig. 12. Fig. 13.

Thus in Fig. 12 the prism is increased by the prismatic action due to decentration of the lens, by shifting the visual axis toward the apex of the constant prism, while in Fig. 13 it is decreased by a decentration in the opposite direction.

Supposing a 5 D. lens to be combined with a prism of 2^{Δ}, the former being decentered 2 m/m, by shifting the visual axis toward the apex of the prism. We know that a 5 D. lens will produce 5^{Δ} when decentered 1cm, and therefore will produce 0.2 of 5^{Δ} when decentered 2 millimeters, which is equal to 1^{Δ}. The constant prism of 2^{Δ} has therefore been increased by 1^{Δ}, making it 3^{Δ}. A decentration of the lens to an equal amount in the opposite direction will leave but 1^{Δ} for the entire combination.

Two millimeters have in this case affected the value of the constant prism by 50% of its active function.

We can now realize the importance and necessity of an accurate adaptation of spectacles, with regard to the inter-pupillary distance, when high spherical corrections are resorted to.

The prism-dioptry and the meter-angle being directly dependent upon the inter-pupillary distance, it behooves us, in any endeavor to secure accurate results, to be exceedingly particular as to its measurement.

Dr. Stevens having fully shown the disturbances occasioned by hyperphoria, we may here be permitted to call attention to the danger of artificially producing it by improperly centering the lenses in the vertical meridian in a simple case of hyperopia in which no real hyperphoria exists.

We shall admit that the lenses are decentered vertically, in opposite directions, by 5 $^m/_m$ above and below the horizontal plane, which will be equivalent to a decentration of one of the lenses by 1 centimeter, provided the other is properly centered.

Our hyperope being corrected by lenses of 2 D., would, under these circumstances, be forced to overcome 2^Δ at the meter-plane, and therefore a vertical diplopia of 12 centimeters at 6 meters, which amounts to 20 meters at one kilometer, or, approximately, in round numbers, 66 feet at a distance of 3,281 feet.

This shows that the vertical adjustment of the lens-centers before the eyes should receive fully as much attention as their horizontal distance apart.

While emmetropes may have one eye much higher than the other, and still enjoy comfortable binocular vision, yet the same discrepancy in ocular elevation in an ametrope whose glasses have been fitted with their optical centers on the same horizontal plane, will produce great discomfort on account of one of the lenses projecting its image eccentrically to the macula of at least one eye. Measurement by the phorometer will in this instance reveal an apparent hyperphoria, which in reality does not exist as a muscular anomaly.

Inversely, in making Dr. Stevens' test* for hyperphoria, in emmetropia for instance, supposing a means to be devised to enable the patient to exactly indicate the distance between the images which he sees at a 6-meter distance. Admitting, by way of illustration, that he has decided them to be 6 centimeters apart, vertically, which, being equivalent to 1^{cm} at 1 meter distance, will lead us at once to decide that a prism of 1^Δ, properly placed before the eye, will correct his *manifest* hyperphoria. The same patient would have to struggle with a vertical diplopia amounting to $3\frac{1}{3}\,^m/_m$ at a

* Functional Nervous Diseases, by George T. Stevens, M.D., Ph.D., New York, 1887, page 194.

reading distance of $33\frac{1}{3}^{cm}$, being quite sufficient to cause consecutive lines of small type to appear intermingled with each other.

Shortly after this paper was first published, the author devised the following simple apparatus (phorometric chart) for estimating the deviations of the visual axes.

The chart here described has, however, been plagiarized by one to whom it had previously been exhibited.

In connection with it the author preferably uses a 12 D. cylindrical lens, which produces a much heavier line of light than the Maddox rod.

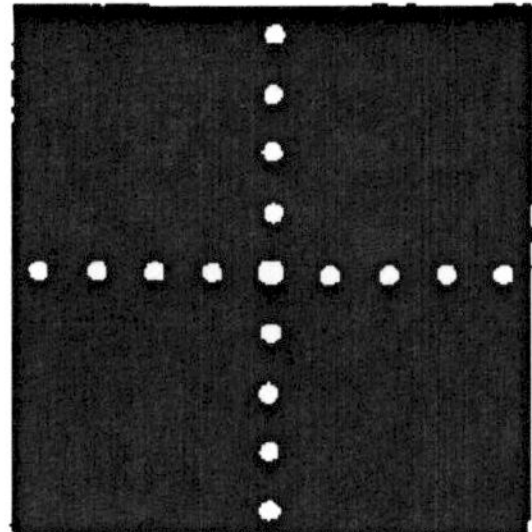

Fig. 14.

Fig. 14 shows a blackboard 20 inches square having on its surface eight vertically and eight horizontally arranged dots, which are separated by 6 sentimeter distances, so that each interval of space between the dots represents 1^{Δ} at 6 meters distance from the eye. A light is placed behind a piece of red glass in a circular opening in the center of the board.

The patient being directed to look through his *distance* glasses—at the central light with both eyes, while the cylindrical lens is held by the operator, first vertically and then horizontally before one eye—will be able to promptly indicate any deviation of the red line of light from the center, by stating through which of the dots the red line seems to pass.

For instance, should the red line of light appear to pass horizontally through the second dot above the center, when the cylindrical lens is properly held by the operator before the patient's right eye, we at once decide that a prism of 2^{Δ}, placed with its base up before the patient's right eye, should cause the red line to drop two points to the center, thus correcting the *manifest* vertical deviation of his visual axes.

Lateral deviations of the visual axes are to be similarly determined by means of the horizontal dots on the board.* Great caution must be exercised, however, in reaching conclusions respecting lateral deviations, as prismatic corrections in such cases are rarely so satisfactory as when prescribed vertically. It is of course of the utmost importance that the centers of the distance glasses worn during these tests should be carefully adjusted in respect to their inter-pupillary distance and elevation.

As a third demonstration, supposing it to be desired to afford binocular vision to an emmetrope at a distance of ⅓ of a meter, *without his powers of accommodation and convergence being called into requisition.* His inter-pupillary distance being $60^{m}/_{m}$, for instance, half of it will be 3 centimeters, which gives us 3^{Δ} for *his* meter-angle, making 9^{Δ} requisite to set aside his convergence to ⅓ of a meter, while 3 D. of lenticular refraction are necessary to substitute accommodation for the same distance. A 3 D. lens of sufficient size would require to be decentered 3^{cm} to afford 9^{Δ}, so that a pair of such lenses, placed before the eyes with their bases in, would accomplish the desired binocular result. Two properly centered lenses of 3 D. combined with prisms of 9^{Δ} would serve the same purpose.

The above illustrations suffice to show the value of the prism-dioptry in leading to our conception of the actual work performed by prisms at different distances, and which the degree-system of numbering must continue to keep us in ignorance of.

Besides, a degree-system of numbering prisms cannot be brought to a *convenient* relation to any of the following considerations, which have been shown to exist in favor of the metric system:

1. A direct relation between the meter-angle and the prism-dioptry for variable inter-pupillary distances.†

---

* When the deviation of the visual axes exceeds the amount provided for by the chart (4^{Δ}), the case may be properly considered to indicate surgical intervention.

† This fully meets the suggestion of Dr. Maddox, who, in speaking of the decision of the committee of the American Ophthalmological Society, consisting of Drs. Edward Jackson, S. M. Burnett, and Henry D. Noyes, that all ophthalmological prisms should be *marked* with the angle by which they deflect rays of light, in his work entitled "Ophthalmological Prisms," on p. 80, says: "If I may be allowed to suggest it, a still better plan would be to have all prisms marked with meter-angles and their fractions, so as to correspond with lenses in the trial case, a meter-angle being the chosen unit of convergence, just as a dioptry is that of accommodation. The only disadvantage is that the meter-angle is an inconstant quantity."

This inconstancy is also mentioned in the first paragraph on page 113 of this paper.

It has, however, been satisfactorily shown that this seeming objection is an advantage to the new system, since the present inconstancy of our prisms can be utilized, thereby securing a degree of accuracy unattainable by any other means.

2. A direct relation between the prism-dioptry and the lens-dioptry, for any amount of lenticular decentration.

3. Simple measurement of the inter-pupillary distance determining the prism which expresses the meter-angle.

4. All fractional intervals of the prism-dioptry being rendered available for differing inter-pupillary distances.

5. The prism-dioptry being capable of measurement by a simple instrument, obviating it being taken for granted that the prisms have been correctly numbered and "marked."

6. The resultant deflection produced by similarly placed superposed prisms of low power being equal to their sum expressed in dioptries.

7. Can be applied to the existing stock of prisms, without increasing the cost, or rejecting the present marketable product.

The latter is certainly a most commendable feature, since any attempt to introduce prisms of special glass, or such that produce definite angular deviations, will heighten the cost, while depreciating the commercial value of those now on hand. However, American manufacturers have now surmounted this obstacle in their production of prisms of the dioptral system.

A possible objection to the new system of measuring might arise in the fact that it does not define the minimum deviation, yet, as the prisms used in spectacles are of small angles, "the difference is so trifling that it may be neglected in ophthalmic practice;"* and, so long as it is understood that prisms of greater angle are to be held with one side parallel with the vertical inter-pupillary plane, from the eye, which is the case in measuring by the prismometer or prismometric scale, we at least obtain the desired uniformity.

In the event of its being desired to determine the resultant prismatic action of prisms which have been combined at any angle of crossing, we have merely to resort to the statical formula:

$$R = \sqrt{P^2 + Q^2 + 2PQ \cos \gamma},$$

where P and Q represent the prisms, expressed in prism-dioptries, and γ the angle between their base-apex lines.

*Ophthalmological Prisms, by Ernest E. Maddox, M.B., London, England, 1889, page 11.

In ophthalmic practice this is apt to prove of possible value only when the prisms cross each other at right angles, consequently when $\gamma = 90°$, and therefore,

$$R = \sqrt{P^2 + Q^2}$$

The resultant prismatic refraction, R, will be in the plane which coincides with the diagonal of the parallelogram obtained by the forces P and Q.

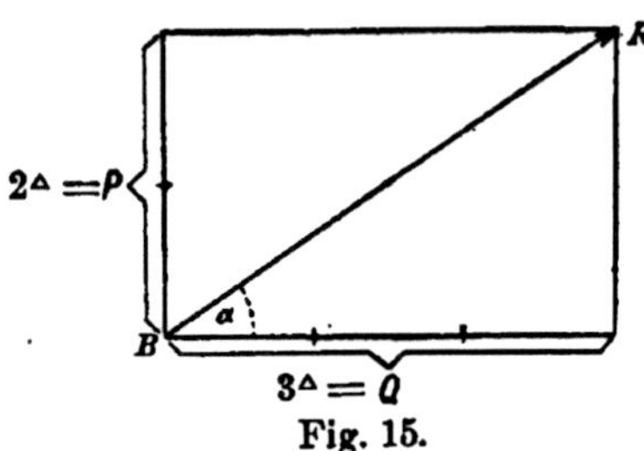

Fig. 15.

This is graphically illustrated in Fig. 15. From a scale of equal parts lay off the values for P and Q (2^{Δ} and 3^{Δ}) perpendicularly to each other, when the length of R, measured by the same scale of equal parts, will represent the power of the equivalent prism in prism-dioptries. The position of the base-apex line of the resultant prism R is obtained by measuring the angle a with a protractor whose center is placed at B.

THE PERFECTED PRISMOMETER:

ITS PRACTICAL ADVANTAGES, CONSTRUCTION, AND VARIOUS APPLICATIONS.

Revised reprint from the "Archives of Ophthalmology," Vol. XX, No. 1, 1891.

In the first numbers of these ARCHIVES of the year 1890, the author described "A Metric System of Numbering and Measuring Prisms," which represented the result of a careful and extensive study of the subject, due to the suggestion of Dr. S. M. Burnett, who had entrusted him with the problem of searching for a system which should prove satisfactory to ophthalmologists as well as avoid conflict with the practical methods of manufacturing opticians. At the close of the author's investigations, he felt that this had not only been accomplished, but that also an instrument in support of that system had been offered as a valuable assistant to opticians.

The author's familiarity with the routine of manufacture would not allow him to lose sight of the practical side, so that this, being a matter of primary importance to opticians, was kept well in view from the outset. In advocating the metric system and the use of the prismometer, we shall therefore here only do so in so far as they relate to the interests of manufacturing and dispensing opticians; the advantages of the system to ophthalmic practice having been previously set forth.

The author's argument in favor of the metric system was, and is based upon the former unavoidable variability in the angles of our prisms, and which must result from the foreign process of manufacture. Although this has been indicated in the previous papers, we shall here take the liberty of quoting from a paper read in connection with the author's exhibit of the prismometer before the New York Academy of Medicine, October 20, 1890:

"It would, however, be exceedingly difficult and correspondingly expensive to manufacture prisms producing only fixed intervals of deflection. To render prisms sufficiently inexpensive as spectacle glasses it is necessary that they should be produced in large quantities at one grinding."

"The process at present consists in fastening a number of slabs of glass, by means of pitch, or other resinous material, upon a metallic surface-tool. The friction in grinding generates more or less heat, which at times is sufficient to soften the pitch and cause it to yield beneath the slabs. Some slabs will shift more than others, so that the prism-angles will vary more or less throughout. Besides, the underlying layer of pitch can never be of a uniform thickness." Were it not for these facts, competition alone would undoubtedly long ere this have resulted in greater uniformity.

By means of the prismometer the author has found prisms, more especially of low degree, to vary between ten per cent. and thirty per cent. of their indicated numbering.

It is obvious that if manufacturers were obliged to discard all those prisms which varied from desired fixed intervals of prism-angles, minimum deviation, or any other designated deflection, the price would have to be increased on the perfect prisms sufficiently to compensate for the cost of those rejected, and which would have consumed equally as much material, time, and labor to produce.

Without being confined to the deflections which should, by calculation, correspond to the prism-angles and index, the author found, by means of the prismometer, among a series of prisms, of best Parisian manufacture, only the following number to produce deflections which were *even* alike:

Three doz. prisms	1°	2°	3°	4°	5°	
Number alike	6 = 1.1	7 = 2	6 = 3.1	6 = 3.7	8 = 4.6	prism-dioptries
Balance varying between	0.8 & 1.6	1.8 & 2.5	2.6 & 3.2	3.4 & 3.9	4.4 & 4.8	" "

These prisms were taken from original packages, and may be credited with having been made of the same material, at the same time, and upon the same tools. Greater precaution on the part of the manufacturer could not be expected.

To the *careful* reader of the author's papers it must have been apparent that *stress* had *nowhere* been laid upon the possibility of a variability in the index, but, on the contrary, that all his deductions were referred to the commonly accepted index of 1.53.

The privileges, however, were mentioned* of which manufacturers might avail themselves, both in respect to prism-angle and index, in seeking to provide prisms of the desired properties.

To any one familiar with the use of optical theodolites† and spectrometers‡ it must further be apparent that an endeavor to measure the *minimum* deviation, with prisms of small angles especially, is very tedious and difficult. The apparatus is expensive, requires a degree of accuracy in manipulation, and a knowledge in the reading of verniers, with which opticians cannot readily be made familiar. To mount such prisms accurately upon the table of the spectrometer, and to rectify the various adjustments of the instrument, are tiresome and slow operations which alone are sufficient to condemn its daily use by opticians whose work must necessarily be expeditious. In the physical laboratory, however, the instrument is undoubtedly invaluable. If the use of an instrument is to be abandoned in measuring the minimum deviation suggested by Dr. Jackson, we shall find that manufacturers will simply divide the prism-angles by two (2), for the new nomenclature, and so give us the old culprit disguised under a new name. There would be great commercial convenience to be sure, in being able to dispose of the same prism under two names, but no *reform* in the interest of scientific exactness could be effected *without* measurement. Will it be policy under such circumstances to adhere to the *minimum* deviation merely for principle's sake? As the prismometer is intended to measure the refraction of prisms, in terms of the prism-dioptry, it may be well, for the benefit of those who may have found its *simplicity* obscured by the mathematical portion of the previous papers, again to explain its principles in more simple and somewhat different terms.

We know that a lens-dioptry is the unit of refraction, and corresponds to a lens of one meter focus, Fig. 1.

---

* Page 109.

† Lehrbuch der Physik., Prof. Joh. Müller, Braunschweig, 1878.

‡ Practical Physics, Glazebrook & Shaw, London, 1889.
Elements of Physical Manipulation, Prof. Ed. C. Pickering, Boston, 1873.

The prism-dioptry, since lenses are but a fusion of prisms of varying angle, may then also be said to be the linear deflection which the refracted ray sustains at the focus of a meter-lens, when the incident ray impinges

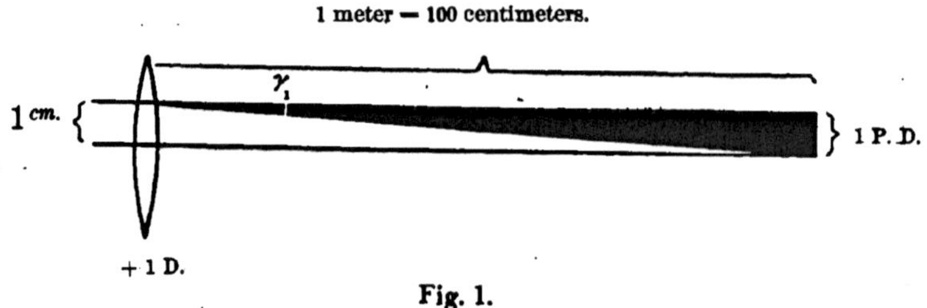

Fig. 1.

upon a peripheral portion of the lens one centimeter from the optical center (Fig. 1).

The prism-dioptry therefore also represents the measure of the angle of deviation γ_1, for an eccentricity or decentration of one centimeter (Fig. 1).

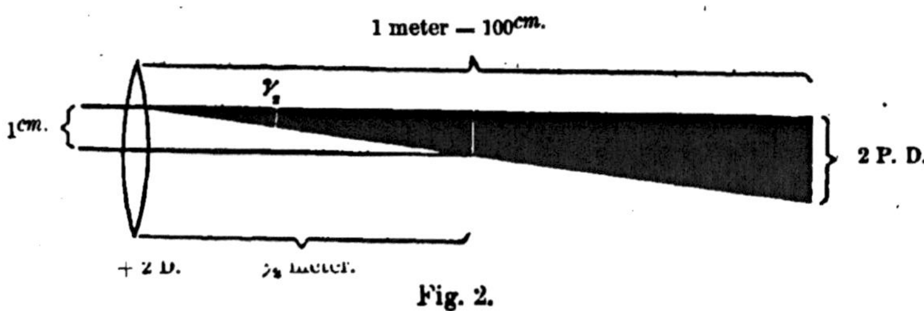

Fig. 2.

A ray impinging upon the same point of a 2-dioptry lens (Fig. 2) will sustain the same unit-deflection at its focus ½ meter, and will therefore find the measure of its angle of deviation γ_2, expressed by twice the deflection at the meter-plane, or 2 prism-dioptries. A lens being decentered twice or half as much will produce twice or half as many prism-dioptries as the lens possesses lenticular dioptries of refraction.* The prism-dioptry is therefore but a sequence to the lens-dioptry. Nothing can be more simple. Thus the prism-dioptry represents the proportion 1:100, which is expressive of a grade of angular inclination in daily use by engineers and scientists the world over. To reduce prism-dioptries to degrees of angular deviation, it is only necessary to divide the prism-dioptries by 100, when they will

* Page 118.

represent the tangents to correlative angles in degrees, which are to be readily found in any table of goniometrical lines. Since different lenses, through varying decentration, will produce different values of the angles of deviation γ_1 γ_2 . . . , how will it be possible to determine the value of such angles in degrees, minutes, and seconds? The instrument is yet to be invented. The prism-dioptry and the prismometer* solve the problem, and in a manner simple and rapid enough to any one of ordinary intelligence.

Since the therapeutic value of prisms is conceded, and their combination with lenses in practice frequent, the prismometer has been constructed with due regard to such combinations, making it possible by its aid to utilize to advantage the prismatic action due to decentration of the lens, for the purpose of offsetting the error which *invariably* exists in the associated prism, after the combination has been ground. Would it not then seem unwise and even arbitrary to hamper the dispensing optician in the practical fulfillment of his work by forcing him to a system of degrees, merely because it harmonizes with the designation of a strabismus which is incorrigible by prisms, or with the graduations found upon perimeters, ophthalmometers, etc., which have no connection with prisms whatever?

The metric system certainly possesses the commendation of reducing all the glasses of the trial case to a uniform nomenclature in dioptries. This alone should be considered a *practical* advantage, fully offsetting the merits of a *theoretical minimum* deviation which cannot at present be *expeditiously* verified by any known means of accurate measurement.

With a view to convenience and simplicity, let us learn to comprehend the power of our prisms by their limits of refraction, shown by the solid triangles in the preceding figures, when it will become wonderfully easy to fit these into meter-angles, or for that matter to any other angles in space, without necessarily confounding prism-dioptries with meter-angles, or meter-angles with "deviations of the eyes in height," as stated by Dr. Landolt.† The latter mistake could only be the result of a misconception of the definition of the prism-dioptry and its relations to the meter-angle.

In recommending the metric system to the profession and practical

* Now also the Prismometric Scale.

† "On the Numbering of Prismatic Glasses," *Archives of Ophth.*, XIX, No. 4, 1890.

opticians, the author in conclusion begs to call attention to its superior advantages, as follows:

1. From a mechanical point of view, by taking the unavoidable difficulties of manufacture into consideration.

2. From a commercial and pecuniary point of view, by avoiding unnecessary expense in the production of prisms.

3. By the prismometer, which enables opticians to accurately fill the demands of the system.*

Any system which neglects these important considerations *cannot* be considered progressive, nor can it effect a *reform* in the present necessarily haphazard endeavors of the dispensing optician, with whom so much of the blame and responsibility must rest. Dispensing opticians have always been on the alert to meet the requirements of the profession, and will no doubt gladly avail themselves of a system and an instrument which will enable them to sustain their reputations as mechanicians.

Taking all the facts into consideration, it suffices to say, that we have prisms of almost every imaginable deflection on hand in the market to-day, so that it merely requires an instrument of simple construction, which may be used in making the proper selection with *accuracy* and *despatch*, and this is precisely what is claimed for the prismometer, which it is the author's purpose here to describe.

In the accompanying illustration, Fig. 3, the essential operative parts of the instrument are shown to be mounted upon a triangular truss which is pivoted by a suitable joint to a pedestal, so as to permit of convenient inclination of the whole.

The graduated bar is rigidly supported near its extremities, upon the truss, by two short studs or pillars, the latter being slightly higher than the radius of the circular stage, which is supported at its back by a rod, fitted, sliding, and acted upon by a spring within the bar, so as to automatically effect contact of the face of the stage with the knife-edge, which is also mounted upon the truss, between the stage and the pinhole eyepiece.

---

*Since this paper was written, American manufacturers have so perfected the art of grinding that they are now able to furnish prisms which are accurately numbered in prism-dioptries, thus causing the prismometric scale to practically supplant the prismometer as an instrument for verifying measurements.

The divisions of the graduated bar, numbered 2, 3, 4, up to 10, are placed at $\frac{1}{2}$, $\frac{1}{3}$, $\frac{1}{4}$, up to $\frac{1}{10}$ of the meter* length, counted from the knife-edge, which represents the zero-end of the scale. A vertical plane, arranged to slide upon the graduated bar, termed the index-plate, is provided with the index-line, marked zero (0), and two graduations at the right-hand upper

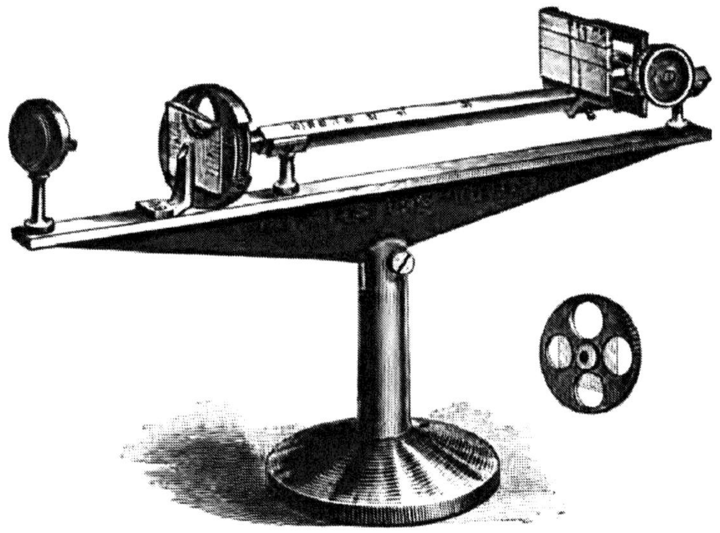

Fig. 3.

edge, marked 1 and 2, which, being equal to correlative centimeter deflections at the meter-plane, correspond to their equivalents in prism-dioptries.

To facilitate subdivision of these graduations the index-plate is provided with a transverse slide bearing its allotted part of the index-line, which is rendered adjustable by a milled head and micrometer screw, the first complete rotation of which will cause this section of the index-line to travel from 0 to 1, the second complete rotation taking it from 1 to 2. The milled head, being divided into 100 parts, enables us, by its graduations, to determine the position of the index-line of the transverse slide, relatively to the graduations upon the face of the index-plate, in 10ths and 100ths.

* It has been found convenient to construct the instrument to half scale throughout.

Thus, in the accompanying figure (4) we read from the face of the index-plate "1" and from the milled head $\frac{2}{10}$ths and $\frac{5}{100}$ths, or 1.25 for the position of the index-line of the transverse slide.

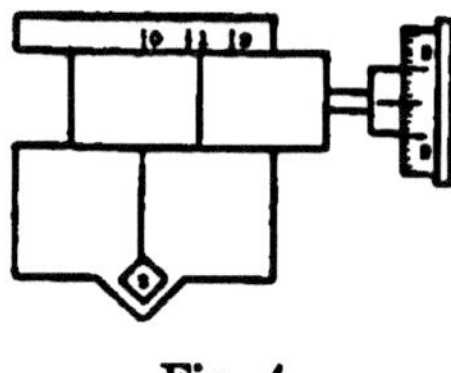

Fig. 4.

As all readings of deflection must be reduced to the *meter-plane*, it will be necessary to note the position of the index-plate, *which must at all times correspond to one of the graduations of the bar.* Consequently, if the above reading is taken from the index-plate, when placed at the figure "2" of the bar, we shall have twice the number of prism-dioptries at the meter-plane, or $1.25 \times 2 = 2.5^{\Delta}$.

For a reading of "2," from the index-plate, when placed at the graduation upon the bar marked "10," we have 20^{Δ}, which is the maximum measuring capacity of the instrument. In other words, it is merely necessary to *multiply* the readings of the index-plate by that figure upon the bar which defines the position of the index-plate upon it.

Before placing a prism in position for measurement, it is necessary to carefully determine its base-apex line. This is accomplished by such slight rotary adjustment of the prism before the eye, until a line, situated at a convenient distance, is sighted as an unbroken one, being precisely the same method which we employ in determining the axes of cylinders. For convenience of registration, ink dots, in collimation with said line, should be applied to the prism. The stage is provided with a series of horizontal lines, engraved upon it, to facilitate perfect adjustment of the base-apex line of the prism, which is to be introduced between the stage and the knife-edge, with its apex to the right, and gradually forced downward while the ink dots pass successively from one horizontal line of the stage to the other, until the upper edge of the prism exactly bisects the circular opening in the stage. In this position the prism will exactly cover the lower half of the opening, while its lateral upper edge will be in collimation with the lower edge of the transverse slide. On completion of this adjustment it is of the utmost importance that the ink dots should coincide with one of the parallel lines of the stage. The observer's eye being placed before the eye-piece, will now perceive the upper edge of the index-plate, and the index-line at zero of the transverse slide, in their true positions, whereas the lower portion of the index-plate, with its index-line, being seen through the prism below, will appear displaced to the right. The position of the observer's eye is now to be carefully maintained, while the graduated milled head is operated with the right hand, until the index-line of the transverse slide has been shifted sufficiently to the right to make contact with the lower index-line seen through the prism. Perfect coincidence of these lines is necessary for an accurate determination of the deflecting power of the prism at any distance. It will consequently be well to previously remove any roughness of the upper base-apex edge of the prism by grinding it to a flat dull edge, and, to be very precise, to take the mean of several readings while the prism is in an undisturbed position. As an example, we shall suppose the prism to have been carefully adjusted in the manner described, and that our readings for three positions upon the bar from the index-plate are as follows:

2d	Graduation	of the	bar,	index-reading		$= 1.57 \times 2 = 3.14^{\Delta}$
3d	"	"	"	"	"	$= 1.05 \times 3 = 3.15^{\Delta}$
4th	"	"	"	"	"	$= 0.78 \times 4 = 3.12^{\Delta}$

$$\text{Mean: } \frac{9.41^{\Delta}}{3} = 3.13^{\Delta} +$$

This precaution, in the interest of exactness, may appear to be unnecessary to some, yet it is here introduced as an exhibit in favor of the capabilities of the instrument.

The prismometer is particularly valuable when it is desired to measure the inherent prismatic action of decentered lenses, and their combinations with prisms.

In such cases it will be necessary to remove a peripheral portion of the lens by grinding it to a dull flat edge, as shown in the accompanying Figure 5.

The lens is then to be placed upon the stage with the flattened edge up, so as to cover half the stage opening; the index-line of the transverse slide having been previously adjusted to zero (0). If, in sighting through the eye-piece, the index-line appears disjoined, it will only be necessary to shift the lens slightly to the right or left to re-establish coincidence of the lines, when the lens is said to be *centered*. While in this position ink dots

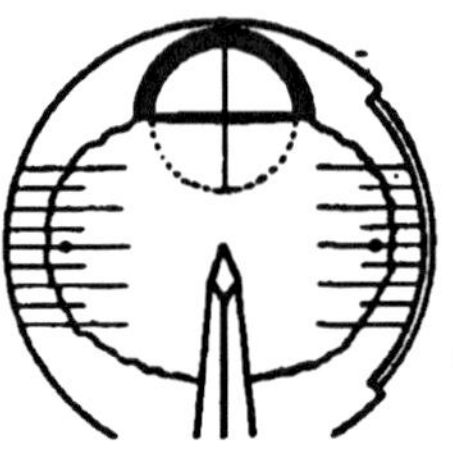

Fig. 5.

should be placed upon the outer edges of the lens over a centrally situated horizontal line of the stage, as shown. For this centered position of the lens, in sighting through the eye-piece, we shall find the index-line at zero (0) unbroken, while the lower half of the index-plate will be enlarged or diminished according to the character of the lens employed. Supposing the lens be 3 D. convex, we shall find the index-plate to present this view (Fig. 6) when it is placed at the graduation marked "3" upon the bar.

Normal Plate.

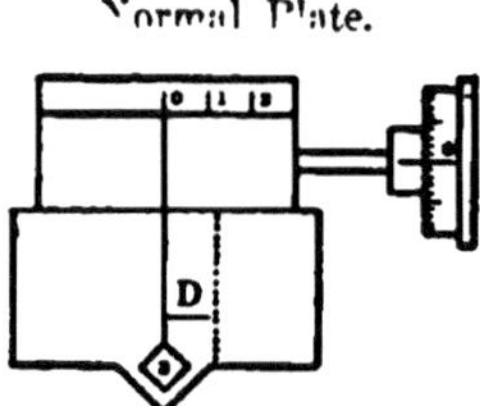

Fig. 6.—Magnified Plate.

The lower half of the index-plate is provided with a red line, indicated by a *dotted* line in the figure, corresponding to a deflection of 1^{Δ}, and which appears proportionately *magnified*. As it will be inadmissible, in our readings, to place a magnified scale on a par with the normal scale of the prismometer, it will be necessay to *register* the magnified unit upon the upper portion of the index-plate, for reference and comparison during decentration of the lens. To accomplish this we displace the index-line of the transverse slide until it coincides with the red line (dotted line, Fig 7), which, as far as the lens is concerned, now *represents* and *takes the place of* 1^{Δ} on the index-plate.

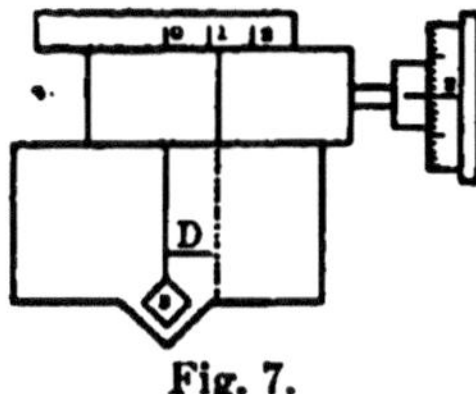

Fig. 7.

Now, by slowly shifting the lens to the left, we shall observe the lines of the lower index-plate to shift to right (Fig. 8).

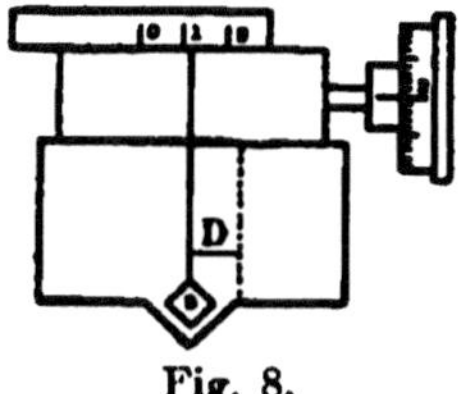

Fig. 8.

When the 3 D. lens has been decentered one centimeter, experiment shows that the lower black index-line cuts the index-line of the transverse slide above. Bearing well in mind that the position of the *upper* index-line now *represents* "1," and that our reading has been taken for a position of the index-plate upon the bar at "3," we have 3^{Δ} as the result of decentering a 3 D. lens one centimeter.*

In case, however, that the refraction of the lens as well as its decentration have not been previously determined it will be necessary to note the following (see Fig. 6):

It is evident that the $\frac{\text{normal plate}}{\text{magnified plate}} = \frac{1}{D}$, so that the

normal prism-dioptries sought $= \frac{1}{D} \times$ the magnified readings, for convex lenses, which will be when $D > 1$, and the

" " " $= \frac{1}{D} \times$ the diminished readings, for concave lenses, which will be when $D < 1$.

* Page 117.

It is therefore only necessary to divide the magnified or diminished readings by D.

The value of D, as we have seen, is determined by first *centering* the lens. It will have a different value for different lenses, and will depend upon the distance of the lens from the index-plate. In fact, D represents the magnifying or diminishing power of lenses for any position of an object, when viewed through them, and which may be placed within their respective focal distances. For the 3 D. convex lens, at the graduation upon the bar marked "3," measurement by the instrument shows D to be equal to 1.2, Fig. 7. Suppose we decenter a + 3 D. lens until we obtain a reading say of 0.6^{Δ}, which is of course a *magnified* reading, we then have $\frac{0.6}{D} = \frac{0.6^{\Delta}}{1.2}$ (magnified reading) $= 0.5$ of the normal prism-dioptry at the distance "3," or 1.5 normal prism-dioptries at the meter plane.

In measuring sphero-prismatic lenses we shall therefore find that the value of the constant prism can either be increased or diminished by a decentration of the lenticular element of the combination, a decentration of 5 m/m in the above instance being sufficient to contribute 1.5^{Δ} ad- or ab-ductive as occasion may demand.

By such means it will be possible to *counteract* the inaccuracies which *invariably* exist in the associated prism after the combination has been ground. When the lens is combined with a prism the flattened dull edge should be cut parallel with the true base-apex line, the latter being registered with ink dots and adjusted upon the stage as usual.

The most ready means of measuring such a combination—for example, + 3 D. spherical combined with 2^{Δ} (constant prism)—will be to place the index-plate at the distance upon the bar marked "3," when, as before, the lens magnification $D = 1.2$, and which may be more conveniently determined by *previously centering* a spherical lens of the same refraction. Now, by deductive reasoning, we know that 2 *normal* prism-dioptries will be equal to $\frac{2}{3}^{\Delta}$ at $\frac{1}{3}$ the distance, and this would require to be 1.2 *greater* at the *same* distance to appear as the properly proportioned magnified deflection seen through the lens, consequently $\frac{2}{3}$ of $1.2 = 0.8$ magnified prism-dioptries. We therefore set the line of the transverse slide so as to read 0.8^{Δ} at the distance marked "3," upon the bar, and proceed to decenter the lens until the lower index-line cuts it, when we shall have the desired

2 *normal* prism-dioptries. We may utilize the rule to prove the result: 0.8 mag. prism-dioptries $\times 3 = \frac{2.4}{D.} = \frac{2.4}{1.2} = 2$ *normal* prism-dioptries. Since lenses are capable of providing as many prism-dioptries as they possess lens-dioptries of refraction, it also follows that we shall occasionally be enabled to secure a considerable proportion of prismatic action by decentration alone, provided the spherical lens is of proportionately greater strength. For instance, the 3 D. lens will produce 3^{Δ} for a decentration of $1^{cm.}$ so that an available decentration of $3\frac{1}{3}{}^{m}/_{m}$ could in itself be relied upon to furnish 1^{Δ} of the 2^{Δ} in the lens forming the subject of our example.

To facilitate measurement of concave sphero-prismatic lenses, the stage is provided with a rotating disk, within, containing three prisms of varying power, with their bases down, and which may be successively carried before the lower half of the opening in the stage as occasion may demand.

The object of these prisms is to counteract the prismatic action in the vertical plane, which would otherwise manifest itself by a confusion of the transverse slide in its contact with the lower portion of the index-plate

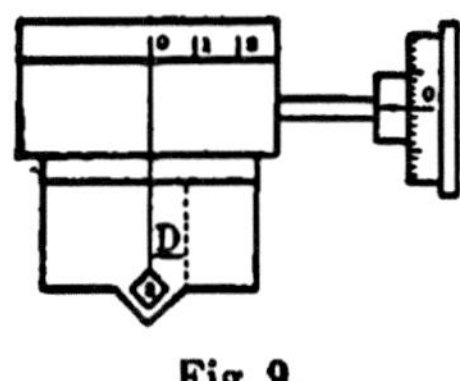

Fig. 9.

(Fig. 9), as a result of sighting through the upper peripheral edge of a concave lens (acting as a prism with its base up) when placed in proper position on the stage. The extent of the confusion of the parts, as shown in the figure, will naturally depend upon the strength of the lens, so that rotation of the disk will reveal the prism best calculated to re-establish contact, as shown in Fig. 10.

Our choice of the prism being made, the lens is to be removed from the stage so as to *rectify* the position of the disk-prism before the index-line at zero (0), which should naturally present a *perfect* vertical line to view.

As an example, let us suppose the combination — 3 D. sph. ◯ 2^{Δ} (constant prism) to be presented for measurement. We should first select a concave 3 D. lens, centering it upon the stage as described, and discover a confusion of the index-plate, at "3" upon the bar, as shown in Fig. 9.

It will be found that the first prism of the disk proves sufficient to re-establish contact, as in Fig. 10.

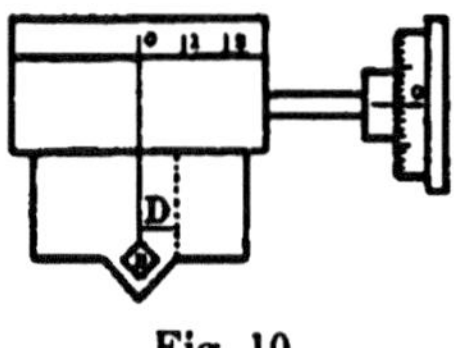

Fig. 10.

Removing the lens, *rectifying* the disk-prism, and *replacing* the lens, we find the diminishing power of the lens $D = 0.83$. Our object being to secure $2^{\triangle}$ at the meter-plane, it follows that $\frac{1}{3}$ of this will have to be the reading from the index-plate at "3" upon the bar, or $\frac{2^{\triangle}}{3} = 0.67$ in the absence of diminishing power, and consequently $0.67 \times 0.83 = 0.55$ as a result of diminution by the lens.

The index-line of the transverse side is therefore to be set to 0.55. The spherical lens is now to be replaced by the sphero-prismatic lens, with its base-apex line marked and adjusted upon one of the horizontals of the stage, and shifted upon this to the right or left, until the lower index-line cuts the index-line of the transverse slide. While the sphero-prismatic lens is in this position, an ink dot is to be placed upon it at the knife-edge, as the dot is intended to ultimately occupy the center of the frame in which the lens is to be mounted.

Such can be the accuracy of the optician's work, with the aid of the prismometer for the metric system, and of which oculists in America may readily avail themselves by a simple request to have their diagnostic prisms re-numbered by measurement upon the instrument. By these explanations the author hopes to have succeeded in conveying the fact, that his object has not only been to promulgate a *theory*, but also to render it *useful* and fully *subservient* to *practice*, and in the absence of which it should, like many another, only live in minds, and mould in books.

THE PRISMOMETRIC SCALE.

Revised reprint from the "American Journal of Ophthalmology," October, 1891.

During the past two years "The Metric System of Numbering and Measuring Prisms"* has been a subject of considerable discussion, although the exact nature of its unit, the prism-dioptry, does not seem to have been generally understood, while its practical advantages to opticians, "of whom accurate work is expected," have been wholly disregarded in some recent criticisms, in which it has been compared with Dr. Jackson's and Dr. Dennet's equally as scientific though less *convenient* systems. It is, therefore, now proposed to call attention to a still more simple feature of the metric system, with further explanations, yet with the understanding that the reader is familiar with its general principle and applications as originally explained.

The prismometric scale, preferably drawn upon heavy paper or card board, consists of a series of numbered gradations, "6 centimeters apart,"

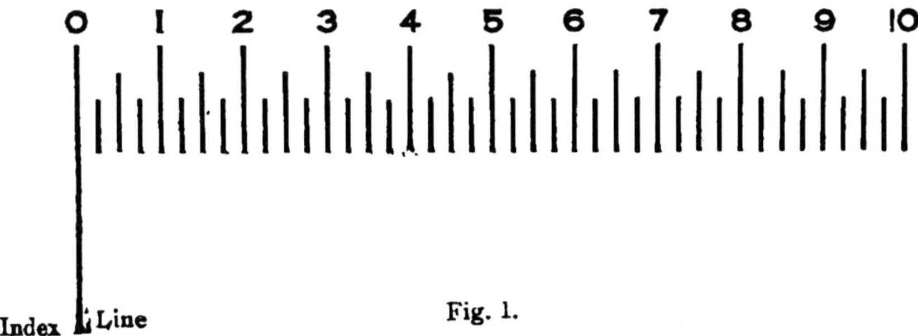

Fig. 1.

with an index-line at zero, longer than the rest, as shown in Fig. 1, which being just six times greater than the "coarse centimeter scale" re-

* See page 105.

ferred to in the author's first paper, is intended to be placed at a six times greater distance, or "6 meters" from the eye; when simple prisms may be measured by it according to the manner originally set forth.

The scale is also subdivided to quarters, thus making possible the measurement of prisms from 0.25 to 10 prism-dioptries.

The *average* deflections produced by our foreign commercial prisms, marked 1° to 5°, will be found to correspond closely to this scale up to the fifth division.

In applying the scale to the measurement of sphero-prismatic lenses, it is evident that the index-line will be rendered more or less indistinct in viewing it through such a lens, so that the lenticular element of the sphero-prismatic lens will require to be fully neutralized by a *contra-generic* lens of the same power, when, by shifting the neutralizing lens from right to left, it will be possible to secure a position for it which will leave us the prismatic deflection which it is sought to attain by the inherent prism of the entire combination.

The procedure is best explained by the following example: The optician being requested to grind a sphero-prismatic lens of + 3 D. sph. ⌣ 2^Δ, selects from his stock a prism which is *rough* on one side, and which he consequently is obliged, from its *marking*, to take for granted is a prism of 2°. He then grinds the rough side to + 3 D. spherical, when, according to the old method, he naturally assumes that he has accomplished the full object of his purpose. It is now suggested that he carefully determine the optical center of a *concave* lens of 3 dioptries, and mark this point with an ink dot, placing the opposite side of this neutralizing lens in contact with the spherical side of the sphero-prismatic lens which it is desired to measure. He is next requested to hold the entire combination before his eye, at exactly 6 meters from the scale, the precaution being taken to have the base-apex line of the sphero-prismatic lens horizontal, with the base to the left, and in such a manner that the upper edge of the entire combination covers only the lower half of the pupil. The index-line observed through the lenses will then appear to be displaced toward the right, relatively to the graduations as seen through the uncovered upper portion of the pupil. In the event of the index-line appearing to be displaced more or less than the required graduation marked "2," the operator has only to shift the neutralizing lens carefully to the left or right, until the index-line exactly cuts the second graduation. Care should be exercised not to change the position of the sphero-prismatic lens during this act, and while in this position, an ink dot should be placed on the sphero-prismatic lens, precisely opposite to the dot on the neutralizing lens. The former then indicates the point which should form the center of the glass in the spectacle frame.

The reasons for this will be obvious from a consideration of the following figures:

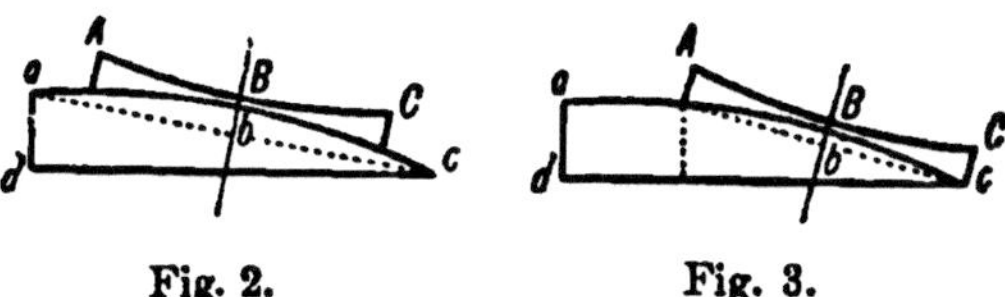

Fig. 2. Fig. 3.

The concave lens ABC in Fig. 2, with its center at B, neutralizes the plano-convex lens *abc*, thus securing the *effect* of a prism *acd*, *just at the opposite points Bb.* By shifting the neutralizing lens, as shown in Fig. 3, the effect of a prism of greater angle is obtained. It is, consequently, possible, within reasonable limits, by this means to correct any inaccuracy which may have existed in the original *rough* prism. The same effect is obtained in sphero-cylindro-prismatic lenses, by neutralizing the cylindrical element with an additional and carefully adjusted *contra-generic* cylindrical lens, though this is naturally a little more difficult. Opticians who keep sphero-cylindrical lenses in stock will generally find it more convenient to use these in neutralizing compound lenses involving prismatic power. It is obvious that it will be much easier to hold and shift a neutralizing lens which consists of only one piece of glass. In shifting the neutralizing lens, great care must be exercised to keep both cylindrical axes parallel in case a change from their coincidence becomes necessary to secure the desired prismatic power.

We shall preface a further discussion of this question with a few simple optical definitions, which the author holds to be indispensable to a thorough understanding of the subject, and which, much to the author's regret, and for reasons too obvious to mention, were not presented by him in the previous papers.

1. The optical center of a lens is a point situated upon a line called the

optical axis, which must be perpendicular to both the anterior and posterior surfaces of the lens.

2. Direct Pencils.—Rays of light which are emitted from a luminous point upon the optical axis will be refracted and directed to a conjugate point upon the same axis, it being specifically noted that the axes of the incident and refracted pencils of light and the optical axis of the lens *must* coincide.

3. Oblique Pencils.—In any case where the axis of the incident cone of light does not coincide with the normals to the surfaces of the refracting medium, whether it be a lens, prism or plate, the *refracted* pencil will no longer be a circular cone of light; but, it will be a pencil bounded by a surface which penetrates and defines the illuminated area of the medium and two focal lines, which are at right angles to each other and the axis of the refracted pencil (see Fig. 6).

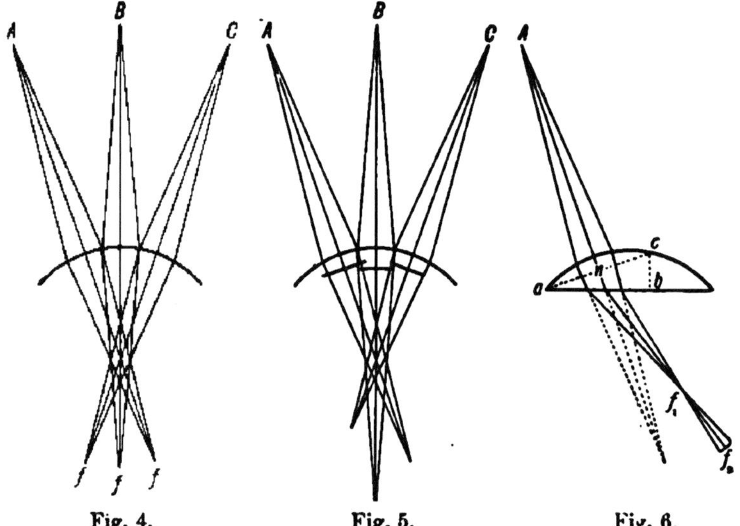

Fig. 4. Fig. 5. Fig. 6.

The same laws apply to the *reflection by spherical surfaces* of direct and oblique incident pencils of light, and their mathematical elucidation is given by Profs. R. S. Heath and W. Steadman Aldis, in their recent exhaustive treatises on Geometrical Optics.

In illustration of the above definitions, let the curved line in Fig. 4 represent the spherical surface of a medium with a greater density than air, when perpendicularly incident conical pencils of light, projected upon it from successive points A, B, C, will have their respective conjugate foci, f, upon the correlative radii with which the axes of the incident pencils coincide. If the refracted pencils, *within* the medium, are to have *focal points outside* of the medium, the axes of these pencils will have to be *perpendicularly* intercepted by the second surfaces as shown by the heavy lines in Fig. 5; and in the event of the second surface occupying an oblique position, *ab*, Fig. 6, with respect to the pencil A, the medium must be considered as a lens, having its optical center upon the axis An of the incident pencil, with the prism *abc* added to it.

The circular cone of light, *within* the medium, will then project an elliptical area of illumination, E, Fig. 7, upon the second surface, as the

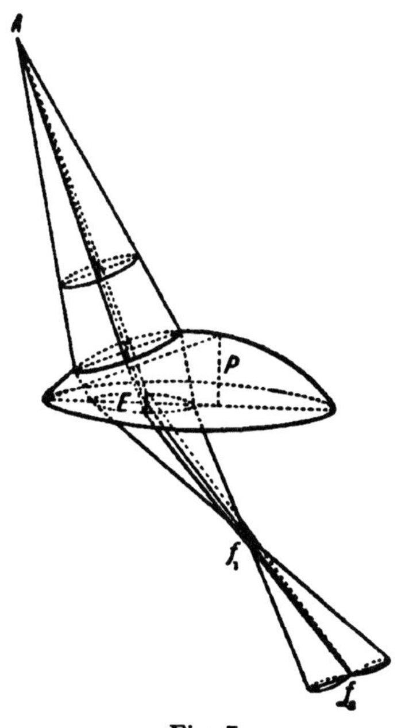

Fig. 7.

axis of the pencil is here *oblique*, and the refracted pencil ceases to be a circular cone, projecting itself outside of the medium as an *astigmatic* pencil, of which f_1 and f_2 are the focal lines at right angles to the axis, the whole being deflected toward the base of the inherent prism P.

While this optical phenomenon, which in this case we may term a sphero-cylindro-prismatic action, may be new to many, it has been known to physical science since Kummer, in 1860, first called attention to the theory by which it was mathematically demonstrated. Its significance to optometrical practice may, perhaps, be treated of in the future.*

The fact, however, may be experimentally, though crudely, demonstrated by placing a plano-convex lens of 8 D. directly between a light at 20 feet, and a screen receiving its image. On interposing a prism of 20^Δ, for example, with its base down, and in a manner to insure contact of the plane faces of the glasses, the image will be observed to change both its form and position upon the screen. By drawing the screen slightly nearer to the lens, a horizontal though imperfectly defined line corresponding to f_1 will become manifest, and by increasing the distance between lens and screen a vertically elongated looped figure, closely resembling a line at f_2, will appear.

When a circular cone of light, *C*, Fig. 8, from a short definite distance falls *obliquely* upon the face of a simple prism, we again have an elliptical area of illumination, and the refracted rays *within* the medium will assume a direction as if emitted from the focal lines v_1, v_2, reaching the second surface of the prism, and being refracted by it to the eye at *E*, as if projected from the lines V_1, V_2, on the opposite side of the prism.

There is one exception to this result, and that is when the axis of the incident pencil assumes a direction which is subject to minimum deviation,

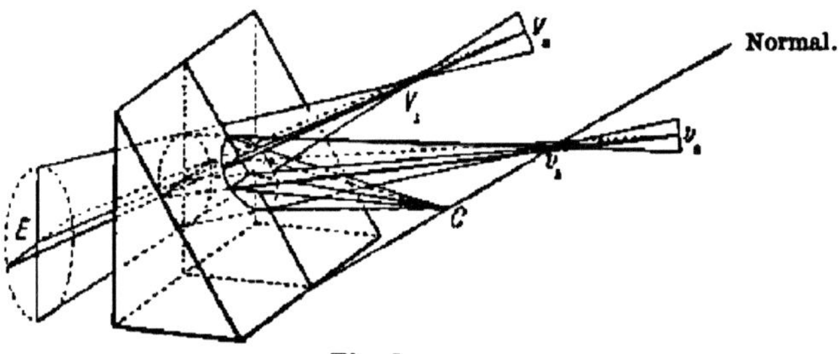

Fig. 8.

in which event the emergent pencil will appear to diverge from a *point*, at the same distance from the anterior surface as the original source of light *C*. In the case of a plate, the emergent pencil will also be of astigmatic

* Now mentioned in Hand Book of Optics, for Students of Ophthalmology, W. W. Suter, M.D., 1899.

form, with the difference that it will appear to proceed from a pair of focal lines located upon an axis *parallel* to the axis of the incident pencil.

This sphero-cylindro-prismatic action, on the part of a simple prism may be experimentally demonstrated in the following manner. Construct the figure *MO* (to the left in Fig. 9, in which the width of the principal lines is, say, 2 inches, and the distance apart of the perpendiculars is

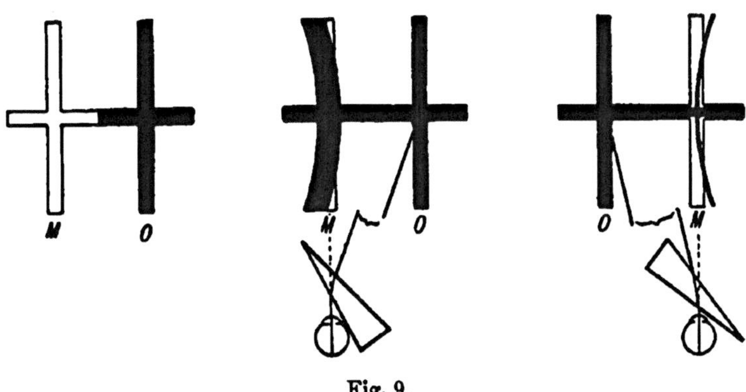

Fig. 9.

about 24 inches), and place it at right angles to the line of sight, at a distance of about 6 feet from the eye, before which a prism of 10^{Δ} is given considerable inclination to the visual axis, with its base in or out, and as shown in the diagrams, to the right in Fig. 9. The eye in each instance is to be placed directly opposite to the figure *M*. In both cases the prism is shown not only to have changed the position of the solid cross *O*, but also to have altered the dimensions of its vertical and horizontal bars in comparison with *M*. With these facts in mind we may return to our subject of measurement.

In Fig. 10 the relative positions of the object of fixation *O*, the prism, and the eye are shown. It is evident that the perpendicularly incident axis *OV* of the conical pencil of rays emitted by the object *O* coincides with the visual axis, and that the axis of the refracted pencil *VP does not* enter the eye, although it *does* define the deflection *OI* which it is desired to ascertain. The axis of the refracted pencil, d_1E, which *does* enter the eye, however, will result from that incident pencil whose axis is *oblique* relatively to the normal at *d*, and it will therefore be a ray approaching direction for minimum deviation and will consequently suffer less deflection, *Oi*, than the refracted pencil whose axis is *VP*.

Now if, as is the case with the prismometer, the observer reads the deflection *Oi* at the definite distance marked, say, "10," upon the graduated

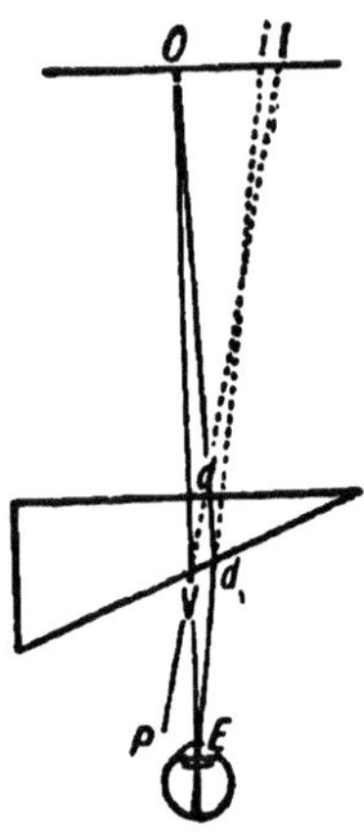

Fig. 10.

bar, it is evident that an error will be committed, since 10 times *Oi* will be less than 10 times *OI*; * yet this seeming weakness in the author's previous papers has escaped detection by the critics of the prism-dioptral system, and for the consolation of whom let it now be said that there could have been no reasoning so clever or ingenious on their part as to have made this error any the less apparent, *even in a prism of 10°*, by merely contrasting the differences between arcs, sines and tangents, in a choice for the unit of measurement. Besides, a mere consideration of the well-known relative goniometrical values of these has not hitherto been pertinent to the discussion, since the proposed unit, the prism-dioptry, is not a goniometrical unit, but an *optical unit*. The desire to multiply any *unit in optics* should be curbed by a knowledge of the fact that all the fundamental optical laws are based upon the assumption and acceptance of *values of limited magnitude*, and that there is therefore apt to be a point where *unreasonable* multiplication of an optical unit will contradict the actually existing optical phenomenon. *A warning to this*

* This will be equally true for measurements taken from an arc at short finite distance.

effect was given in speaking of the decentration of lenses (see page 116 of the author's second paper).

Even *thickness*, a dimension which we are taught to neglect with respect to ophthalmic lenses, becomes an appreciable factor in prisms above 8^{Δ}, when we attempt to measure their deflection at short finite distance. This will be apparent from the following considerations.

It has been shown that the ray, which in the nearest limit reaches the eye, is the axis Od, Fig. 11, of an *oblique* pencil, being refracted within the prism ABC from d to d_1, and thence in air to the eye E, which projects it to i, upon the scale OI. For a given thickness of prism, this is the *only* pencil which will be received by the eye, since, if we increase the thickness by allowing the plane A_1B_1 to represent the anterior surface of the prism, the original incident axis Od will be refracted at v instead of d, when the axis of the refracted pencil will traverse the path vv_1P_1 to the left of the eye and parallel to dd_1E. The refracted pencil which would enter the eye, for the indicated *increased* thickness, could only accrue from an *increased* obliquity of the incident axis Ou. The latter would therefore even more closely approach position for minimum deviation, from which we are to conclude that the deflection noted upon the scale OI by the eye will be *least near the base* and consequently greatest near the apex of the prism.

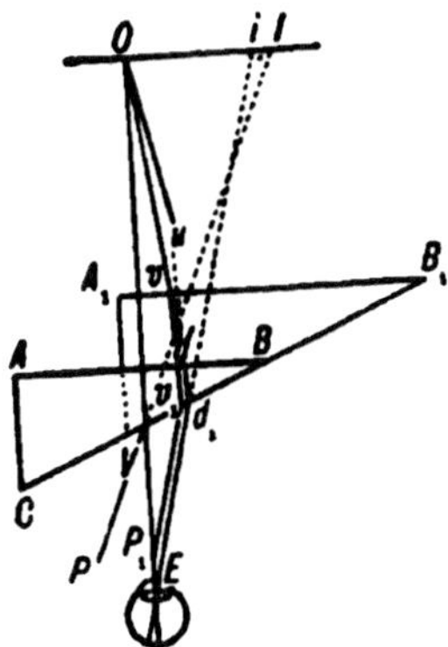

Fig. 11.

This is really proven to be the case by experiment with the prismometer. At the distance marked "10" upon the bar, the index-plate-reading near the base of a 22° prism, 1¼ inches square, is found to be 1.79, whereas at the feather-edged apex it is 1.89, so that the prism in the former instance measures 17.9^Δ^, while in the latter it is 18.9^Δ^. The same prism measured by the prismometric scale at 6 meters reads 20^Δ^. The error committed by measurement through the apex, at short finite distance, is therefore 1.1^Δ^, while the increased thickness at the base still further increases the error by 1^Δ^. The error will consequently be least, *in prisms of high degree*, when readings on the prismometer are taken at the apex of the prism, and it will be reduced to a minimum, *throughout the principal refracting plane*, when the deflection is measured for pencils which are *perpendicularly incident to all points of the prism-surface*, that is to say, when the pencils of light are *cylindrical*, and which will practically be the case when the object of fixation, *a line*, is situated at 6 meters distance. In fact it will be better to measure all prisms above 8^Δ^ at this distance.

This sharply defines both Dr. Burnett's and the author's reason for advocating the tangent plane for the position of the scale, since it will be infinitely more *convenient* to place such a scale upon a flat wall, with which every office and workshop is provided, than to *contrive* an arc of 6 meters radius.

Other advantages of the scale at a 6 meter distance were mentioned in the author's second paper, when referring to hyperphoria.

The above facts do not lessen the value of the prismometer, which the author has repeatedly and specifically represented as being of importance to opticians in filling oculists' prescriptions, in which the prisms do not exceed 5^Δ^, and by reason of which the error is so slight as to be inappreciable, yea, even in a prism as high as 8^Δ^, when an attempt is made to verify measurement by the prismometric scale at 6 meters.

It was also to be supposed that all oculists and opticians would not provide themselves with prismometers, in which event it was further anticipated that the prismometric scale would have to be resorted to, and more particularly now that hair-splitting fractions of the unit are not considered to be of value.

A more simple and convenient means of verifying the opticians' work could certainly not be placed in the hands of the oculist.

The prism-dioptry does not exclusively depend upon trigonometrical laws, nor rest solely upon the adoption of a specific instrument, but it is based upon a principle which is easily understood and capable of being practically applied within the confining limits set by the fundamental laws of optical science. It must also be apparent that the generally irrelevant criticisms which have appeared in print have not, thus far, proven anything to the contrary; while it must be equally clear that this paper contains a review of the optical laws and phenomena which must be considered in the choice of a unit, and that these will require to be thoroughly understood, before anyone can undertake a rational criticism of the subject. We can, therefore, only admit that a perpetuance of the old degree system, together with the commonly accepted approximations which must accompany its application in practice, will serve no better purpose than to obviate such intelligent pains being taken.

ON THE PRACTICAL EXECUTION OF OPHTHALMIC PRESCRIPTIONS INVOLVING PRISMS.

Revised reprint from the American Journal of Ophthalmology, January, 1895.

It is intended here to point out some instances in which lenticular decentration may be taken advantage of in the execution of prescriptions involving a combination of prisms with spherical and cylindrical lenses, in a manner to insure absolute accuracy, with least inconvenience of measurement, and with minimum expense of production. Within the past few years the practical value of the prism-dioptry, as a unit of prismatic power, has been appreciated to such an extent that the American Optical Company of Southbridge, Mass., and the Bausch & Lomb Optical Company, of Rochester, N. Y., have entirely discarded the old degree system of numbering prisms, having now supplanted it by the prism-dioptry for their entire product. The prism-dioptry is therefore no longer a subject for scientific discussion, but one for practical consideration, having been indorsed in its underlying principle by both the American Ophthalmological Society and American Medical Association, and by two of the most progressive and largest manufacturing establishments in the world.* It may be of interest to note that the gross productions of these firms amounted to over $2,000,000 in 1892. Such practical support certainly portends an enduring future for the prism-dioptry, about which so much *pro and con* has been written since its first appearance in ophthalmic literature. In this paper it will be taken for granted that the reader is at least familiar with the principle of the prism-dioptry, so that only the relation which exists between it and the lens-dioptry will here be repeated, to wit:

* See foot notes, pages 101 and 102.

A lens which is decentered one centimeter will produce as many prism-dioptries as the lens has dioptries of refraction.

A knowledge of this law will frequently enable the optical practitioner to change the form of his prescription so that it may safely be entrusted for execution to any optician capable of locating the optical center of a lens, and who may be provided with no other instrument of measure than a pocket centimeter scale.

As the dispensing optician, generally speaking, is not allowed to exercise his own judgment in transforming prescriptions, it is necessary that the oculist's instructions should be explicit respecting the proper method of putting his prescription into the best practical form. A few examples will serve to illustrate the method of applying the law of decentration.

Prescription No. 1.

O. D. + 3 D. sph. ◯ 1^{Δ} base out.

O. S. + 3 D. sph. ◯ 1^{Δ} base out.

The usual practice is to grind + 3 D. sph. upon a prism of 1^{Δ}, so that a plano-convex spherical element of 3 D. is substituted for the bi-convex lens used in the trial frame. In another paper, "The Advantages of the Sphero-Toric Lens," attention is called to the positive disadvantages of this procedure, especially where high degrees of curvature are concerned.

However, in the above example the ordinary method of grinding a spherical surface upon one of the faces of a constant prism is also objectionable on account of the increased cost, so that for two very important reasons it is always preferable to resort to decentration of the lenses, where that is possible, than to grind sphero-prismatic combinations. The aforesaid prescription shows that the prism-dioptries required are few compared to the lens-dioptries, so that the law of decentration becomes available. In accordance with this law, 3 D. sph. decentered 1 cm. gives 3^{Δ}, and since only 1^{Δ} is needed, a decentration of $\frac{1}{3}$ cm. for each lens will satisfy the requirements. To avoid unnecessary expense, the prescription should therefore be written:

O. U. + 3 D. sph. decentered $\frac{1}{3}$ cm. toward the temples, by which we mean that the thick edge of each lens should be placed at the temples, as in Fig. 1, which is the right eye.

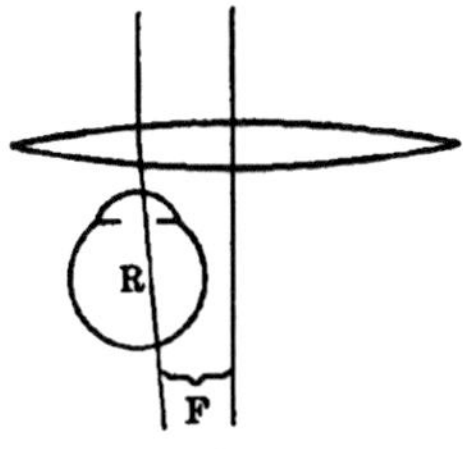

Fig. 1.

The optician's ability to apply the law of decentration is hampered only by one obstacle at present, namely, by the limited size of the lenses now furnished by the manufacturers.

These lenses, as they are usually supplied to the trade, are capable of a decentration of only $\frac{1}{3}$ cm. laterally, and $\frac{1}{2}$ cm. vertically. But even under these circumstances there are many prescriptions in which decentration may be profitably utilized. For instance:

Prescription No. 2.

O. D. — 2 D. cyl. 180 ⌒ 2$^{\Delta}$ base up.
O. S. — 2 D. cyl. 180.

Might be written:

O. D. — 2 D. cyl. 180 decentered $\frac{1}{2}$ cm. down.
O. S. — 2 D. cyl. 180 decentered $\frac{1}{2}$ cm. up.

The decentration of $\frac{1}{2}$ cm. on a 2 D. cylinder produces 1$^{\Delta}$, so that 1$^{\Delta}$ base up on the right eye, and 1$^{\Delta}$ base down on the left is equivalent to 2$^{\Delta}$ base up on the right alone. The term "decentration" signifies a displacement of the lens-center, relatively to the visual axis; hence a decentration of $\frac{1}{2}$ cm. "down," on the part of the axis of the concave cylinder, means that the axis is displaced downward, thereby placing the thick edge of the lens up, as in Fig. 2. The decentration would of course have to be in the opposite direction for a convex cylinder to produce prismatic refraction with the base up. Whenever prismatic corrections in the vertical direction are necessary, a preference should be given to place the base up before one eye rather than base down before the other, especially where the prism is stronger than 2^{Δ}. This is explained in the fact that the eyes are much more frequently turned downward than upward, and are therefore more exposed to the annoying internal reflections, which are noticeable near the base of the prism, when its base is also down.

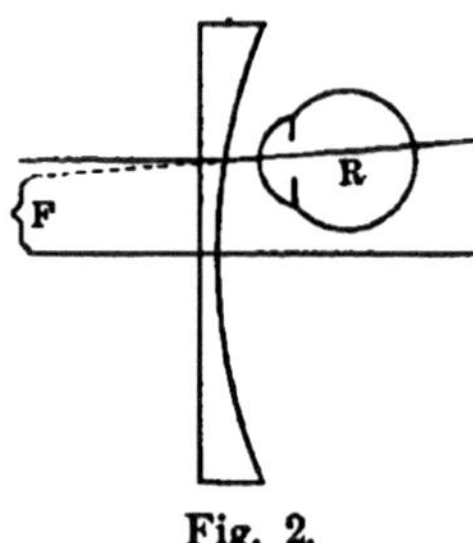

Fig. 2.

This, in part, also explains why prisms with their bases towards the nose frequently fail to prove so satisfactory as they otherwise might.

As another example let us cite a case in which the examination results in

PRESCRIPTION NO. 3.

O. D. + 4 D. sph. ◯ 2^{Δ} base out.

O. S. + 2.75 D. sph. ◯ + 1.25 D. cyl. 180 ◯ 2^{Δ} base up.

In this instance the lens of the right eye would require to be decentered ½ cm. "out" to secure 2^{Δ}, since 4 D. decentered 1 cm. gives 4^{Δ}. We know, however, that ½ cm. decentration is in excess of the amount (⅓ cm.) generally available in a lateral direction.

But, if we turn our attention to the left lens, we see that it is possible to call upon the 2.75 D. spherical to supply a part of the lateral prismatic action, as the necessary displacement can be made along the cylinder's axis without implicating the cylinder. In other words there are 6.75 D. of spherical power before both eyes which may be called into requisition to provide an equivalent to 2^Δ base out for one eye.

The 6.75 D., decentered 1 cm., gives 6.75^Δ, or decentered one millimeter $= \frac{1}{10}$ of $6.75 = 0.675^\Delta$. As 2^Δ are needed, it will take $3 \times 0.675 = 2.025^\Delta$, that is to say it will take a decentration of 3 millimeters on the part of each lens toward the temples to satisfy our requirements.

The left eye calls for 2^Δ base up. We have vertically 2.75 D. sph. + 1.25 D. cyl. 180, equivalent to 4 D. of refraction, which is capable of producing 2^Δ by a vertical decentration of $\frac{1}{2}$ cm. = 5 millemeters. The prescription should therefore read :

O. D. + 4 D. sph. decentered 3 millimeters out.

O. S. + 2.75 D. sph. ◯ + 1.25 D. 180 decentered 3 millimeters out and 5 millimeters up.

These examples suffice to show how easily and with what absolute accuracy these prescriptions may be executed without incurring the additional expense of grinding prismatic combinations. This expense should only be incurred in those cases where decentration is impossible on account of an insufficient size of the lenses. A glance at the prescription will determine at once which of the methods to apply. Take, for examble, a case like

Prescription No. 4.

O. D. + 1 D. sph. ◯ 1^Δ base out.

O. S. + 1 D. cyl. 90° ◯ 1^Δ base out.

In this case, as our commercial lenses are too small to bear a decentration of 1 cm., it would be necessary to grind the lenses as the prescription is written, though even here, to lessen the expense of the left lens, it is preferable to write:

O. D. + 1 D. sph. ◯ 2^Δ base out.

O. S. + 1 D. cyl. 90°.

In this prescription care should be taken to match the lenses as nearly as possible in thickness.

It would be a great convenience to have the optical manufacturers furnish a series of lenses capable of a decentration of at least 1 cm. Such lenses would not require to be larger than 6 cm. in diameter, and could be confined to the weaker powers, say from 0.25 to 2 dioptries. The cost of such lenses should certainly not be greater than that of sphero-prisms, and would offer many opportunities of applying the prism-dioptry expeditiously, with greater accuracy, and less inconvenience to the dispensing optician.

A PROBLEM IN CEMENTED BI-FOCAL LENSES SOLVED BY THE PRISM-DIOPTRY.

Revised Reprint from Annals of Ophthalmology and Otology, January, 1895.

It is intended here to illustrate the principal defect which so frequently leads to disappointment in the use of cemented bi-focal lenses, as well as to explain the technical means by which it may be prevented. When occasion demands, it is common practice among oculists to prescribe glasses for "reading" and "distance," with rather vague instructions to the optician to provide the necessary lenticular corrections in the form of bi-focal lenses. These, in the event of their being of the so-called "cemented" variety, the optician executes by cementing wafers of glass to the surfaces of the lenses which fit the areas enclosed by the eye-wire; both of the wafers being cut from the peripheral edges of that lens which produces the requisite amplifying or reducing power in the lenticular combination. The principal effort of the optician, at present, is to make this lens as thin as

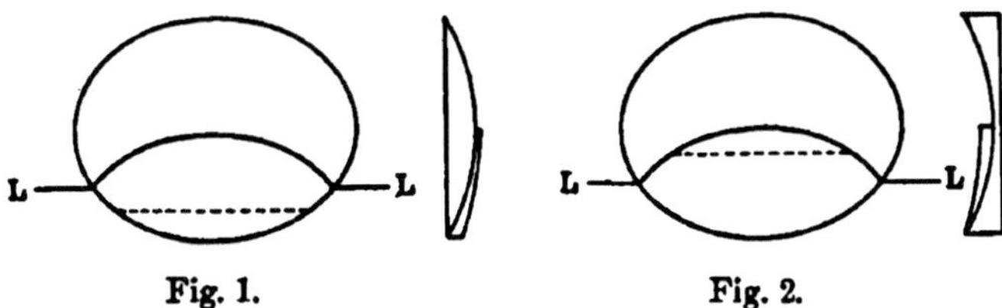

Fig. 1. Fig. 2.

possible, and to reduce its diameter so as to enable him to secure at least two wafers of sufficient size for the ocular fields required.

Economy and extreme thinness of wafer are doubtless desirable, but these are only of minor importance. Spectacles as now constructed, exclusively with this in view, are rarely ever free from a prismatic action, operating vertically, which renders them very uncomfortable to wear, and frequently useless, or harmful. This is especially true for high degrees of curvature, *and in cases involving combination with cylinders*, where the spherical refraction is obtained by spherical curvature of *one* surface only. With a view to brevity, only the latter type of correction will be discussed.

The accompanying diagrams, Fig. 1 and Fig. 2, will serve to illustrate the defect referred to.

In each of these the line *L* to *L* is drawn upon the paper which is supposed to be placed several inches behind the bi-focal sphero-cylindrical lens. The line viewed through the cemented lens appears disconnected, being deflected by the prismatic action resulting from decentration of the "distance" and "reading" lenses, relatively to each other. It is, of course, customary to have the lower smaller field for reading, as above shown, but, for convenience of easier demonstration, the reader may make the interesting experiment of superposing two concave lenses, say — 4.5 D. and — 2.5 D., on which the optical centers have been previously marked with ink dots, and allowing them to occupy the positions shown in Fig. 3, in which the overlapping parts are in the smaller field for distance.

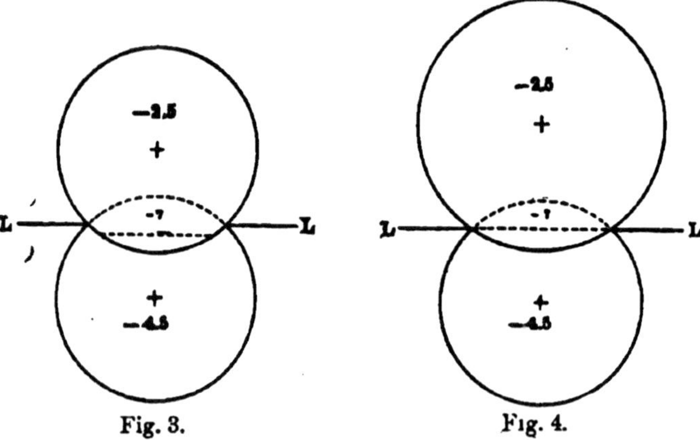

Fig. 3. Fig. 4.

By more widely separating the lens centers (+), it will be observed that the disconnected portion of the line *LL*, as seen through the lenses, will appear displaced to a lesser degree, and, if the upper concave lens 2.5 D.

is chosen of sufficient diameter, it will be possible to secure a distance between the lens-centers which will exhibit the line *LL*, unbroken, as in Fig. 4. This shows that the prismatic action depends only upon the distance separating the lens-centers, and therefore also that *the lens*, specifically in this case the upper one, from which the segmental wafers should be cut, *must have a definite diameter for every combination*, if the prismatic action is to be eliminated. We will cite an example in which we have for distance: — 7 D. sph. — 2 D. cyl. ax. 180, and for reading: — 4.5 D. sph. — 2 D. cyl. ax. 180, which, executed as a cemented bi-focal lens, calls for a + 2.5 D., periscopic wafer. This leads us to the proposition:

What peripheral segment of a 2.5 D. periscopic convex lens should be used as a wafer to insure freedom from prismatic action in the center of the wafer 7 millimeters below the center of the distance lens: — 7 D. sph. — 2 D. cyl. ax. 180?

The key to its solution is to be found in the law that "a lens decentered one centimeter will produce as many prism-dioptries as the lens has dioptries of refraction."*

The center of the wafer being 7 millimeters (0.7 cm.) below the center of the distance lens, makes it obvious that we have a prismatic action at this point acting vertically, on account of the — 7 D. sph. and — 2 D. cyl. ax. 180, which is equivalent to a decentration of 0.7 cm. on 9 D. to be neutralized by the wafer. Reverting to the law we find that 9 D. decentered 1 cm. affords 9^{Δ}, therefore, 0.7 cm. will give 0.7 of 9, or 6.3^{Δ} as the prismatic action to be overcome.

The wafer of + 2.5 D. decentrated 1 cm. gives only 2.5^{Δ}, so that it takes a decentration of 2.5 cm. to produce $2.5 \times 2.5^{\Delta} = 6.25^{\Delta}$. Therefore, this wafer when placed with its thin edge at the lower edge of the concave distance lens will neutralize the existing 6.3^{Δ} with an error of only 0.05^{Δ}. As will be later shown the + 2.5 D. lens, so as to be large enough for so great a decentration, must be at least 64 millimeters in diameter, and should be ground to a knife-edge to insure maximum thinness. As this example came to the author's notice, the instructions given to the

* See page 117.

mechanic, who successfully executed the lenses, are here repeated as follows:

"Make a 2.5 D. periscope convex lens (+ 7 D. — 4.5 D.,) 64 millimeters in diameter, worked to a knife-edge at the periphery, and, after mark-

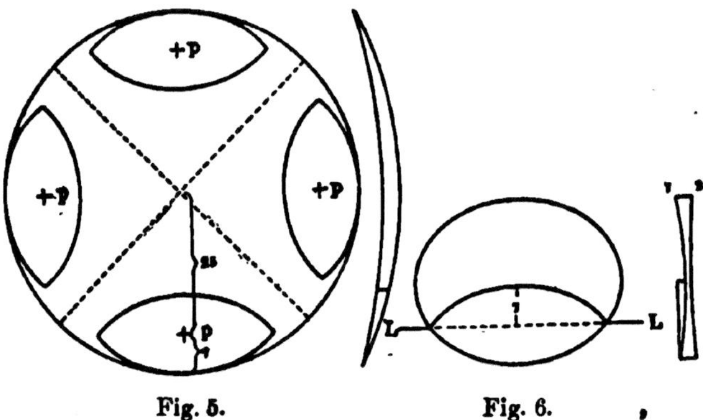

Fig. 5. Fig. 6.

ing its optical center, lay off four points, *p. p. p. p.*, 25 millimeters from the center, as shown in diagram Fig. 5:

"Replace the lens on the *convex* grinding tool, and cut the lens through the indicated dotted right-angled diameters into four equal parts. This will secure four quadrants, with ample provisions in case of accident. Select for both eyes two of the most perfect quadrants, and cut from them the peripheral segments to the shape indicated, and cement the wafers into the concave 7 D. spheres, with their thin edges down, when a line *LL* viewed through the reading lenses will appear continuous as in Fig. 6."

It is obvious that the diameter of the lens is determined by adding twice the decentration (25 × 2) to one full width of the wafer, (7 × 2) which gives 64 millimeters.

It would be very difficult to solve this simple problem so easily by any other means than the prism-dioptry.

WHY STRONG CONTRA-GENERIC LENSES* OF EQUAL POWER FAIL TO NEUTRALIZE EACH OTHER.

Revised reprint from *Annales d'Oculistique* (English Ed.) November, 1895.

Some time ago Mr. George W. Wells, President of the American Optical Company, Southbridge, Mass., requested the author to give this subject attention, and, as it contains features of mutual interest to oculists and opticians, it has been considered of sufficient importance to give the results of this investigation publicity.

In practice it is customary to determine the power of a lens by what is known as "neutralization." It is here proposed to show why it is that this method is only strictly applicable to lenses which are weaker than 9 dioptries. The power of a lens, as is well known, is dependent upon three factors—the radius of curvature, index of refraction, and thickness of glass. The latter we are taught to consider a negligible quantity, since it is generally infinitely small in proportion to the focal distances of lenses which are used in spectacles. This is only justified in its application to *concave* lenses, since all concave lenses, between 0.25 and 20 D., can be made of the same infinite thinness in the center. In convex lenses, however, we meet with an unavoidable increase in thickness, which becomes of sufficient magnitude in lenses above 8 D. to conflict with the hypothesis referred to. When the element of thickness is considered, we have the formula for bi-spherical lenses of equal curvatures:

$$r = F(n-1) + \sqrt{\frac{(nF-e)\,F(n-1)^2}{n}}, \quad \ldots \ldots \quad \text{I.}$$

* Lenses of opposite character—convex and concave.

wherein r is the radius of curvature, n the index of refraction, F the focus, and e the thickness, in contra-distinction to the formula for neglected thickness, wherein

$$r = 2F(n - 1). \quad \ldots \ldots \ldots \ldots \quad \text{II.}$$

It is therefore evident that the radius of curvature will be a different one for bi-convex lenses, in which thickness is considered, from that of bi-concave lenses, of the same power, having no appreciable thickness.

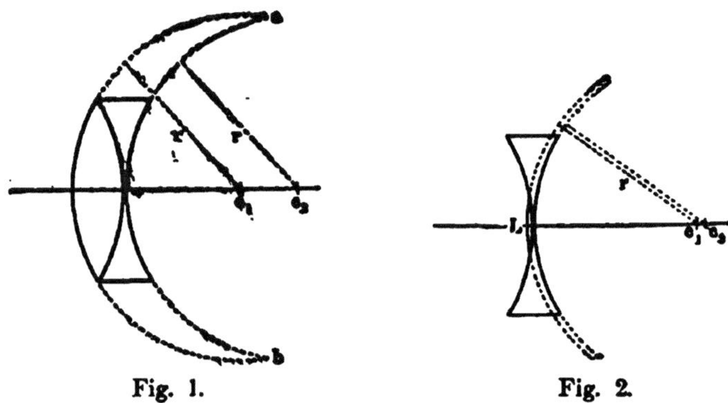

Fig. 1. Fig. 2.

The accompanying diagram, Fig. 1, representing a bi-convex and a bi-concave lens of *identical* curvatures, clearly shows that they cannot optically neutralize each other, as they really only constitute the central portion of a much larger periscopic *convex* lens, and which our imagination can construct upon the dotted lines which are continued to their intersections at *a* and *b*.

The diagram also shows that the power of the imaginary convex meniscus must *increase* with an increase in the thickness of the bi-convex lens, because the anterior and posterior surfaces of the meniscus will be rendered more oblique to each other as their respective centers of curvature, c_1 and c_2, are separated to provide for an increased thickness. The nearest approach to neutralization will therefore be secured when the centers of curvature, c_1 and c_2, are as close together as possible, thus making the bi-convex lens L exceedingly small and thin, as shown in Fig. 2.

The lenses in our trial cases are, however, too large to secure even this *approximate* neutralization. Their diameter of necessity determines the thickness, which must increase with the power. For instance, in a 20 D. convex trial-case lens of 3.5 centimeters diameter we find the minimum thickness to be 0.75 centimeters. If, therefore, in Formula I, we place $e = 0.75$, $n = 1.5$, and $F = 5$ centimeters, we obtain 4.87 centimeters as the value of r, whereas, for a 20 D. concave lens, according to Formula II, we find $r = 5$ centimeters.

As the radius is shorter for the convex than for the concave lens of the same power, it is evident that their surfaces will actually only touch in the center, as exaggeratedly shown in Fig. 3.

Besides, the outer surface, s_1 s_1, of the convex lens will be even more oblique relatively to the outer one, s_2 s_2, of the concave lens, so that these lenses actually form the center of a stronger convex meniscus than shown

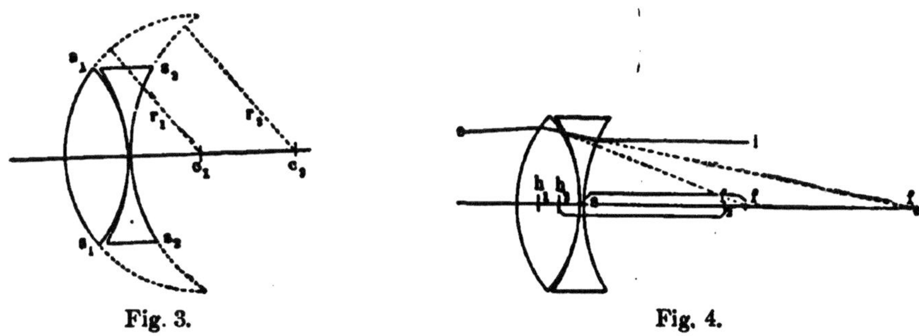

Fig. 3. Fig. 4.

in Fig. 1. Or, viewing it in another light: Parallel rays, i, which are incident to the concave lens, are refracted by it, passing into the convex lens, as if emanating from the virtual focal point f, of the concave lens, which is outside of the focal point, f_2, of the convex lens in Fig. 4.

Such rays, e, will therefore be rendered convergent, instead of parallel, in passing out of the convex lens, showing that neutralization does not exist.

The focal distance, in infinitely thin lenses, is counted from a single point, s, on the optical axis in the center of the lens, whereas in lenses of appreciable thickness it is counted from the ***posterior principal point***, h_2, within the lens. In a bi-convex lens of equal curvatures, made of glass, with an index of refraction $n = 1.5$, it has been demonstrated that the principal points, h_1 and h_2, are separated by a distance equal to one-third of the lens thickness.* It is therefore obvious that the focal distance of the convex lens will have to be increased by at least one-third of the lens thickness, so as to have f_1 and f *coincide* for the purpose of effecting neutralization.

With a minimum thickness equal to 0.75 cm., we must, consequently, add 0.25 to the focal distance, 5, making 5.25 the focal distance of the convex lens. This corresponds to a refraction of 19.047 dioptries. Consequently, *a 19.047 convex lens of 0.75 cm. thickness neutralizes a 20 D. concave lens of no thickness.* To be more accurate, we should actually allow for one millimeter thickness of the concave lens. This would result in the convex lens being even somewhat weaker than 19.047 dioptries.

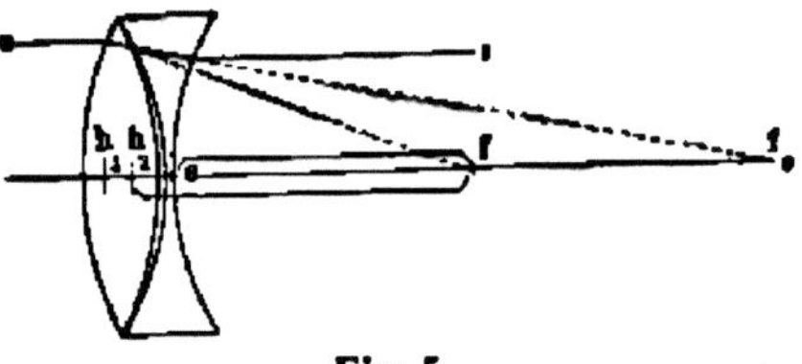

Fig 5.

However, according to Formula I, a 19.047 D. convex lens of 0.75 cm. thickness should have a radius $r = 5.121$ cm., so that the superposed neutralizing lenses would actually touch each other at their edges, instead of at the center, as exaggeratedly shown in Fig. 5.

Furthermore, any additional increase in the thickness of the convex lens will be associated with an increase in the distance $h_2 f$, and will therefore call for a corresponding decrease in the power of the convex lens to produce neutralization.

Thus, the tendency will invariably be to overestimate the power of the convex lens, when an effort is made to determine its power by that of the concave lens which neutralizes it.

Calculation shows that discrepancies in neutralization varying between 0.25 and 1 D. exist in the entire series of convex lenses between 9 and 20 D. It consequently also follows that the indiscriminate addition of lenses, as frequently practiced during the subjective method of examination of ocular refraction, is not permissible for lenses of high power.

*Müller-Pouillet's *Lehrbuch der Physik*, page 160. Braunschweig, 1894.

In other words, lenses of high power are no more capable of being algebraically combined than prisms of high power, a fault for which the practicability of the prism-dioptry was so severely criticised. The same logic therefore applies to lenses which was mentioned in the author's paper on the Prismometric Scale, to wit:

"The desire to multiply any *unit in optics* should be curbed by a knowledge of the fact that all the fundamental optical laws are based upon the assumption and acceptance of *values of limited magnitude*, and that there is therefore apt to be a point where *unreasonable* multiplication of an optical unit will contradict the actually existing optical phenomenon."

The general, though erroneous, impression that the entire series of corresponding contra-generic lenses should neutralize has gained such credence that lens manufacturers have allowed themselves to be swayed by this popular opinion, and as a result have adopted the principle of making the convex weaker than the concave lenses, so as to meet the demand for neutralization. As has been shown, a 20 D. convex lens should have a shorter radius than a concave one of the same power, yet examination of any trial case will reveal the fact that the reverse is the case, when the surfaces are measured by a suitable gauge. We can, however, gain no reliable information regarding the power of a strong convex lens by

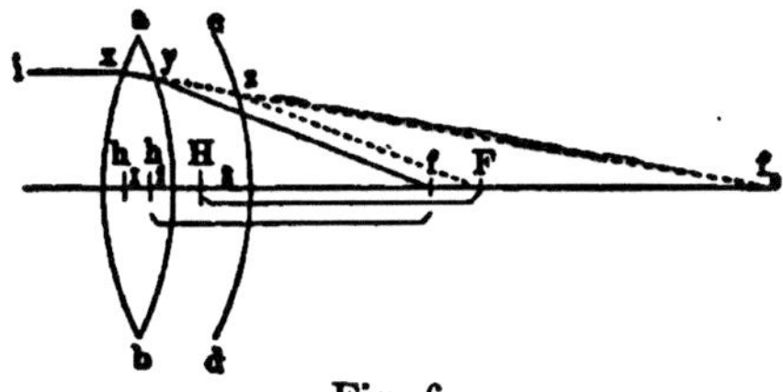

Fig. 6.

measurement of its surfaces, since two lenses of the same curvature, but of different thickness, will not be of the same power. As the thickness increases, the power will diminish for one and the same curvature.

This is shown in Fig. 6.

The incident ray, i, is refracted by the anterior surface of the lens ab in the direction xf_0, and by the posterior surface at y to the focus f. If the thickness be increased, so as to place the posterior surface at cd, then xf_0 will be refracted by the posterior surface at z to F, parallel to yf, since the surface cd is of the same radius as ab.

The focal distance H_2F, in the diagram, is not appreciably different from h_2f; in fact, the difference between these distances scarcely amounts to 0.05 cm., for a curvature of 5 cm. in lenses of 0.75 and 1 cm. thickness. Nevertheless, such a difference would be appreciated by the eye in neutralizing. The question then arises: By what method should we determine the power of strong lenses? Indeed, nothing seems to remain but to measure them by a focusing screen upon a graduated bar, care being taken to count the focal distance from the posterior principal point within the lens. Such precaution is, however, rarely if ever taken. While concave lenses can be similarly measured, yet this is somewhat difficult, and not entirely satisfactory. The lens surface-measure is indeed preferable *only for concave lenses*, since they are of negligible thickness. For convex lenses above 8 dioptries this instrument is absolutely unreliable.

It will, however, be more or less inconvenient to have individually different methods of measurement for convex and concave lenses. Unless, therefore, we are prepared to make some radical change, we shall do well to adhere to the practice of neutralization, bearing in mind the errors we commit by so doing. When convex lenses stronger than 8 dioptries are prescribed, and they are measured by neutralization with a concave lens, we should remember that they are always weaker than the dioptries indicated by the concave lens. In short, *the convex lenses in our trial cases are not what they are numbered in dioptries.*

So long, however, as manufacturers are agreed that the convex lenses they produce shall neutralize with standard concave lenses,* we shall at least have a uniformity which will insure the convex lenses in spectacles being duplicates of those in the trial case; *provided*, also, that the lenses of any given number are always of the same uniform thickness. This thickness, in every instance, should be the *least* which can be given to the lens of 3.5 cm. standard diameter.

* Lenses made by the American Optical Company are ground upon this principle and should be tested accordingly. In neutralizing always hold the convex lens next to the eye. If lenses are made upon the opposite principle, they will not neutralize with these.

THE ADVANTAGES OF THE SPHERO-TORIC LENS.

Revised reprint from the Ophthalmic Record, July, 1895.

The sphero-toric lens having been described in the author's treatise on Ophthalmic Lenses, we shall here proceed to show the various advantages which this lens possesses over its sphero-cylindrical equivalent in correcting astigmatism. This elucidation will, however, be confined chiefly to applications of the toric lens in cases of aphakia.

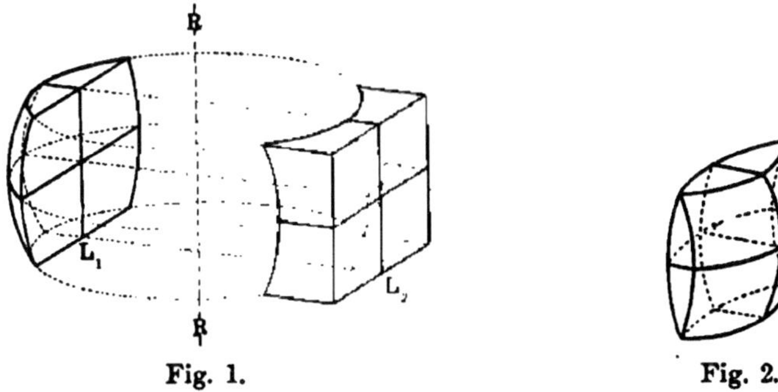

Fig. 1.

Fig. 2.

In Fig. 1, L_1 represents a plano-convex, and L_2, a plano-concave toric lens. Fig. 2 represents a sphero-toric lens with the toric surface in front, and the spherical surface behind.

Similarly, in examining the refraction in aphakia, it is customary, as in other cases, to place the spherical lens in the groove at the back of the trial-frame and therefore nearer to the eye, with the cylinder in front. For example, in a case involving

$$+ 9 \text{ D. sph.} \bigcirc + 3.5 \text{ D. cyl. ax. } 160°,$$

the result is obtained by placing a bi-convex spherical lens of + 9 D. in the trial-frame behind the cylinder + 3.5 D.

The optician, however, carries out this prescription *literally*, by making the lens with + 9 D. spherical on one side, and + 3.5 D. cylinder on the other, thereby substituting a *plano-convex* spherical element for the bi-convex one used in the trial frame. Furthermore, in mounting the lens, he places the spherical surface farther from the eye, with the cylinder inside, so that the spectacles are worn with the surfaces of the lenses *reversed* in respect to their positions before the eye in the trial-frame.

It is, therefore, obvious that the serious error is committed of substituting a lens which involves a change in the position of the "nodal points" of the entire refracting system, thus to a degree impairing the visual acuity obtained by the spectacles, as compared with that secured by the test in the trial frame. Aside from this we have, in the aforesaid sphero-cylindrical lens, the unpleasant phenomenon of internal reflection, caused by the extreme bulging forward of the convex element, whose surface is exposed to light coming from other directions than that of the desired central incident beam.

Every ray of light, *R*, Fig. 3, entering the convex spherical surface in a direction coincident with its radius of curvature, will enter the lens *unrefracted*, and fall upon the inner surface *C* of the cylinder on the other side.

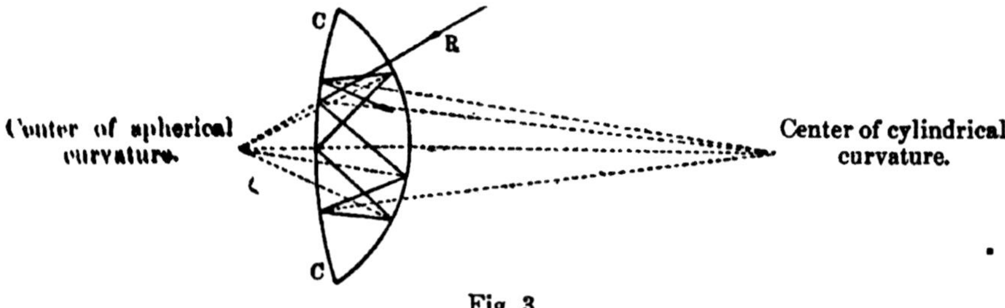

Fig. 3.

Under favorable conditions it will be reflected from this back to the inner anterior surface of the lens, there again undergoing reflection, and so on, until it becomes dissipated within the lens as diffuse light. The dotted lines in the diagram represent the radii of curvature, dividing the angles of incidence and reflection, which are equal at the points of impact on the surfaces within the lens.

The patient wearing such spectacles is therefore compelled to look, as it were, through a self-luminous medium, which tends to interfere with the direct central incident beam of light passing through to his pupil. Again, in this form of lens the aberration is greater for all excursions of the eye where the visual axis passes through the lens at points not coinciding exactly with the optical axis of the lens.

These facts were clearly exemplified in a case which came to the author's notice in 1889. The lens had been made of the sphero-cylindrical form and power as above stated. The patient complained of inability to correctly judge distances in looking down, and of an annoying glare of light which he felt was within the lens itself when worn out of doors, or in a strong light. He thought, if the glare could be removed he would be enabled to see better. In fact he had made the experiment of shading the lens by sighting through the partially closed hand held closely before it, and had noticed less inconvenience from the glare.

The sphero-cylindrical correction, then worn by him, gave him scarcely better than 6/9 of vision, whereas the test lenses in the trial frame, by the author's test, gave him the somewhat unusual acuteness of 6/6.

It occurred to the author that the interior reflections could be avoided by constructing the lens with less curvature on the anterior surface. Therefore, a lens having a toric surface on the anterior side, and a spherical surface on the other was suggested as follows : + 6 D. sph. combined with a toric surface of + 6.5 D. ax. 160° by + 3 D. ax. 70°, in place of + 9 D. sph. $\bigcirc$ + 3.5 D. cyl. ax. 160°.

Before proceeding to consider the equivalence of these lenses, it is suggested that all thought of a simple cylindrical form should be dispelled from the mind when the toric surface is referred to. The subject will be more easily understood by dealing only with the principal refracting meridians, without regard to the other meridians of curvature, which give form to the toric surface.

Referring to the lens, + 9 D. sph. $\bigcirc$ + 3.5 D. cyl., it is evident that the meridian of least refraction is 9 D., and the meridian of greatest refraction is 12.5 D., see *m* and *M* in the perpendicular planes of Fig. 4.

The extreme spherical curvature, 9 D., on the anterior surface may be lessened to any desired degree, provided the loss of refraction is compensated for by adding it to the opposite cylindrical side of the lens, in which case the simple cylindrical surface would necessarily be *replaced* by a surface of astigmatic refraction, a toric surface, having the required focal interval of 3.5 D. To obtain such an equivalent lens of ***thinnest and best form***, it is only necessary to divide the original meridian of greatest refraction, $M = 12.5$ D., in two parts, as nearly equal as possible, say, 6 D,

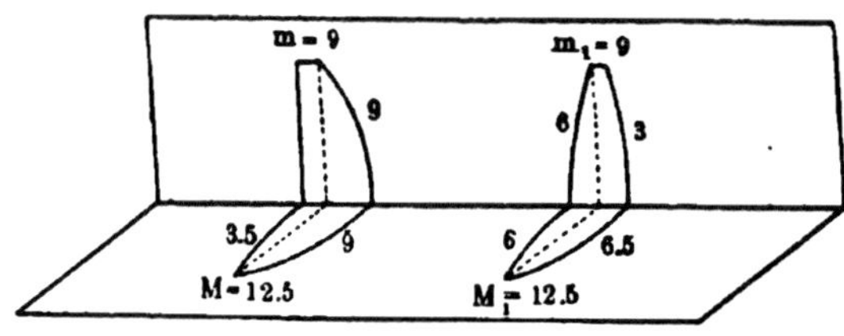

Fig. 4.

for the back surface, and 6.5 D. for the front surface, which, together, give 12.5 D. for the newly-created *bi-convex* meridian of greatest refraction. M_1, Fig. 4, in the desired equivalent lens.

Taking 6 D. as the posterior spherical surface, it is necessary to combine it with 3 D. on the anterior surface to secure 9 D. as the meridian of least refraction, m_1, in the new lens. Consequently 6.5 D. and 3 D., respectively, represent the refraction of each of the principal meridians of the anterior toric surface, which, when combined with the 6 D. posterior spherical surface, will fulfil all the requirements of equivalence as shown in the diagrams.

The sphero-toric lens referred to was set in a spectacle-frame with the toric surface outward, and its meridian of least refraction at 160°, so that the refracting elements occupied the same positions as the lenses in the trial frame.

The spectacles have been worn by the patient for the past five years, with complete relief from all the disagreeable phenomena mentioned. It may also be stated that his visual acuteness with the new lens is slightly better than 6/6.

For reading, the patient required + 12 D. sph. ◯ + 3.5 D. cyl. 160° in the trial-frame, which was given him in the form of a sphero-toric lens: + 8 D. sph. ◯ toric surface of + 7.5 D. ax. 160° by + 4 D. ax. 70°.

In addition to the advantages mentioned, the patient is saved the annoyance of wearing uncomfortably heavy lenses, which also attract attention. In every instance where the author has applied the sphero-toric lens it has given entire satisfaction, though in none of the cases has the visual acuteness been so perfect as in the one just cited.

The use of the sphero-toric lens is by no means confined to cases of aphakia, since equally good results can be secured by its use in high degrees of compound myopic astigmatism, especially where the cylindrical corrections are weak in comparison to the high spherical curvatures involved.

Furthermore, the sphero-toric lens is also capable of being given a periscopic form, and which, if desired, may be made quite as globular as the so-called coquille glass. In this form, especially, it affords the advantage, when placed before the eye, of allowing its peripheral area to be brought nearer to and more concentric with the eye-ball than is possible with the sphero-cylindrical equivalent; so that, for all ordinary motions of the eye, the visual axis will be less oblique to the inner surface of the lens. This, and the consequent absence of reflection from the inner concave surface, also give this form of sphero-toric lens a wider and more natural field of vision than is obtained by the ordinary sphero-cylindrical equivalent. This feature is appreciated to a marked degree by those who wear them in shooting, or at billiards, tennis, golf, etc.

A further commendable feature is that sphero-toric lenses of high power can be made very much thinner, and consequently of less weight than is ordinarily possible in such cases. They are also frequently advantageous where cemented bi-focal lenses are required, since the segments for reading are inclined at an angle more closely approaching position for perpendicular incidence of the visual axes when looking downward.

Although sphero-toric lenses are considerably more expensive than others, it is fair to predict a decided increase in their application by competent optical practitioners, as soon as the aforesaid advantages shall have become more widely known.

THE IRIS, AS DIAPHRAGM AND PHOTOSTAT*

Revised reprint from the "Annals of Ophthalmology and Otology," October, 1895.

Under this title it is proposed to inquire into the value of sub-decimals of the lens-dioptry in ametropia. The subject has been frequently discussed, and again, at considerable length, at a meeting of the American Medical Association, in San Francisco, Cal., 1894, with the result that "low degree lenses" are now generally conceded to have a noteworthy therapeutic effect; though no scientific reason has been given, and simply because the physical laws involved have never even been mentioned. While unanimity of opinion of this sort may be exceedingly satisfactory from a medical point of view, yet it only circumstantially corroborates that relevant scientific argument which should properly also embrace the following important considerations.

In every compound lens-system we are met with the necessity of providing against spherical aberration. This is accomplished, in the construction of optical instruments, by introducing an annular disk, of *calculated* diameter, known as the *diaphragm*, which is suitably placed between the lenses to exclude peripheral rays. If the proper diaphragm be replaced by one of smaller aperture, we increase the definition, but diminish the extent of field and illumination. A larger aperture will increase illumination and field, but definition will be impaired, on account of the aberration thus allowed.

The aperture of the diaphragm must therefore have a definite and specific diameter for every optical instrument, if we are to secure *maximum* definition and illumination, *without aberration*. The proper diaphragm is

* Photostat, Greek, φῶς(φωτ-), light, + στατός, verbal adjective of ἱστάναι, stand — an automatic light regulator (suggested by the author).

therefore one of the most important and indispensable parts of every compound dioptric system. The human eye is such a system, and is provided with its diaphragm—the iris. In the eye, which is a dynamic apparatus given to variations of power, a fixed diameter of pupil would fail to theoretically fulfil the requirements. When the eye is in a state of accommodation, it becomes a stronger refracting system, and therefore needs a smaller aperture of diaphragm, hence the pupil contracts.* Yet, Helmholtz† says: "A. von Graefe observed in an eye from which he had removed the iris by operation that the normal range of accommodation was still present, and also that the changes in the anterior curvature of the lens could still be observed." He concludes: "The iris does, therefore, not play an important rôle in accommodation." (Lit. trans.) Landolt‡ expresses the same opinion. So far as the above noted measurements are concerned, such conclusion may be quite correct, yet if construed in its broadest sense it discountenances the value of the iris as a diaphragm entirely.

It is, nevertheless, universally admitted that the iris does act independently of, and simultaneously with accommodation. § When acting independently of accommodation, the iris is known to behave as a highly sensitive *photostat*, through regulating the volume of light upon the retina to such a degree as shall be most agreeable to our light-perceptive sense.

A most subtile and synchronous balance, between retinal perception, uveal stimulus, and iritic response, must therefore exist, if the iris is to perform its functions simultaneously as diaphragm and photostat.

An endeavor will here be made to support the hypothesis that *a disturbed equilibrium of these functions is probably the cause of asthenopia in low degrees of ametropia.* From a strictly optical point of view every eye of the same refraction, other things being equal, should have a pupil of the same diameter—one suited, by *calculation*, to exclude peripheral aberration, while securing the greatest tolerable illumination. This, however, is

*In fact, it was at one time supposed that contraction of the pupil was the only means by which the eye adapted itself for near vision. Helmholtz, "Physiologische Optik," page 151, Hamburg and Leipzig, 1886.

† Helmholtz, "Physiologische Optik," page 138.

‡ Landoldt, "Refraction and Accommodation of the Eye," page 164, Philadelphia, 1886.

§ "Movements of the iris are nevertheless associated with accommodation; they are governed by the same nerves as the latter, so that, until the mechanism of accommodation is better understood, a direct relation between them may not be looked upon as being improbable." (Lit. trans.) Donders, "Refraction and Accommodation," page 485, Wien, 1866.

not known to be the case, nor has the author found that any one has ever calculated what the diameter of the pupil should be for any given schematic eye. Listing has calculated a table showing the changes in diameter of the diffusion circles upon the retina which arise through efforts of accommodation in a schematic eye having a pupil of 4 *mm.**

We have thus far been content to know that pupils differ in size in different persons. There must, however, be a *limit* to the maximum diameter of the pupil, if aberration is to be excluded, and if, *for any reason*, the pupil is prevented from contracting to at least this limit, we shall have aberration, even in the emmetropic eye.

This is *exaggeratedly* shown in Fig. 1, in which the central incident rays, *cc*, focus at *f* upon the retina, while the peripheral rays, *pp*, produce

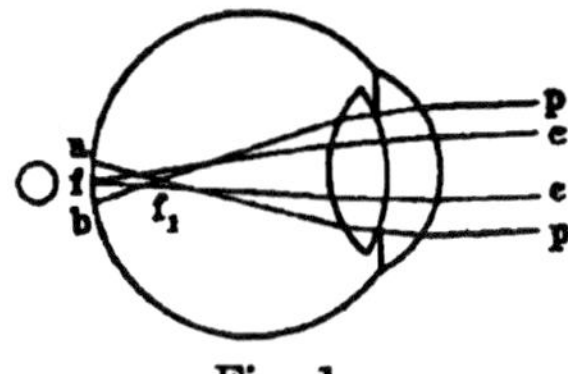

Fig. 1.

thereon an area of diffusion,† *ab*, and which, to all practical purposes, would be equally as effective in impairing vision as a low degree of myopia, having its intra-ocular focus anywhere between the retina and f_1. In fact, it is questionable whether the eye can discriminate between images which are impaired by peripheral aberration and those which are illy defined through slight errors of refraction. The following experiment will serve to illustrate this: By placing a 1 D. convex lens before the emmetropic eye, it is practically rendered myopic for distance, the letters of the test-card at 6 m. becoming indistinct, with a probable reduction in the visual acuteness to, say, $\frac{6}{9}$. If the lens be now covered with a pin-hole disk, normal acuteness of vision will be re-established, with no other appreciable difference than that the field and illumination are less. We may therefore consider the peripheral rays, here accompanying the increased refraction, as aberrative rays in respect to the enclosed central incident beam, so that an eye

* Helmholtz, "Physiologische Optik," page 127.

† For purposes of lucid illustration, the diffusion areas in all of the diagrams are greatly exaggerated.

capable of contracting its pupil to the same extent would, in part, similarly correct its error of refraction.

This is undoubtedly one reason why errors of refraction of the same degree are not accompanied by the same diminution of visual acuteness. The myope of 1 D., with small pupils, *without* glasses, will probably have better vision than the myope of 1 D. with much larger pupils. Within certain limits, peripheral aberration and anomalies of refraction are analogous in destroying definition of the image. A slight error of refraction, with large pupils, may produce diffusion images equally as pronounced as a considerable refractive error with small pupils.

Asthenopia is therefore quite as apt to be experienced on account of the size of the pupil as it is on account of the error of refraction.

This should explain why it is that many persons, having small pupils, endure a considerable error of refraction without inconvenience, whereas others, with large pupils and small errors of refraction, are afflicted with asthenopia.

Again reverting to Fig. 1, the larger the pupil the greater will be the zone of peripheral aberration and its correlated diffusion-area, *ab*. In fact "the peripheral aberration upon the optical axis is known to increase, not only in proportion to the square of the aperture, but, also *pari passu* with the refraction" (physical law), so that we should have greater diffusion circles upon the retina, when the ciliary muscle is brought into action, even in emmetropia, to correct the peripheral aberration which impairs the sharp definition at *f*. The only stimulus which could assist in correcting the aberration in this case would be that which, imparted to the iris from the retina, would cause the pupil to contract sufficiently to exclude peripheral rays. In here speaking of the retina, we of course take for granted its highest state of physiological development. The question then arises: Is such retinal stimulus imparted to the iris in low degrees of ametropia, *independent of accommodation, without increased light intensity?* If there is such independent action on the part of the iris, ineffectual efforts of the ciliary muscle to correct impaired vision may be followed by a contraction of the pupil necessary to shut out the peripheral rays. As to this, let us investigate the relation which should exist between the iris and accommodation in the hyperopic eye.

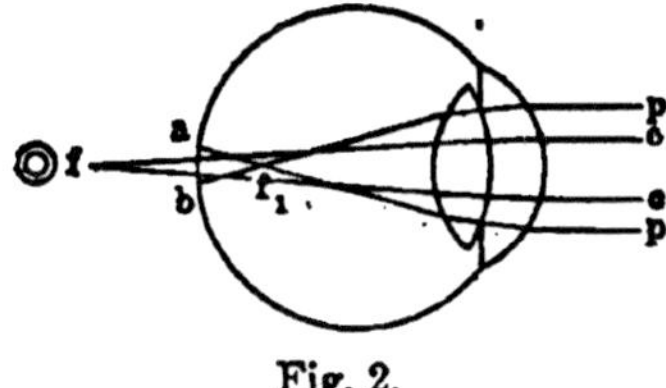

Fig. 2.

In this, Fig. 2, the central rays, *cc*, are focused behind the retina at *f*, the peripheral rays crossing at f_1 and producing the diffusion-area *ab*. In facultative hyperopia there will be accommodation sufficient to bring *f* forward to the retina. With this increased refraction, however, the pupil remaining the same, f_1 will recede from the retina, with a corresponding increase in the size of the diffusion-area *ab*.* It is therefore evident that,

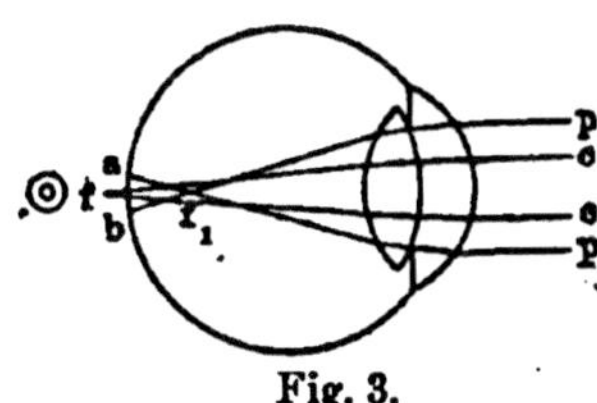

Fig. 3.

if *increased* aberration is to be avoided, a *normal* pupil must contract concurrently with the accommodation. This, generally speaking, is known to be the case. If, as in Fig. 3, the hyperopia is of low degree, with *excessively* large pupil, we shall have a comparatively small central area of diffusion, due to the refractive error, covered by a much larger area of diffusion and illumination, *ab*. The slightest effort of accommodation would tend to sustain or increase this discrepancy. It therefore follows, if the aberration is to be abolished, that the iris must receive an increased stimulus to bring about a contraction of the pupil, *in excess of that which is concurrently associated with accommodation, and that, too, for every degree of*

* Listing's table shows that the diffusion circles upon the retina increase more rapidly as the object approaches the eye at short range. Helmholtz, "Physiologische Optik" page 128.

light intensity. Were this not the case, vision at a distance,* with excessively large pupils, would be impaired by aberration under all circumstances.

The additional stimulus to contraction in undoubtedly due to the *increased area of illumination* above mentioned. This would seem to imply that the contraction of the pupil not only responds to the light intensity (quality) but also to its area (quantity) upon the retina.

It is also evident that the impairment of vision should be ascribed to that factor causing the largest area of diffusion upon the retina. The larger the pupil, the more will the peripheral aberration predominate over that which is produced in the center by a low degree of refractive error.

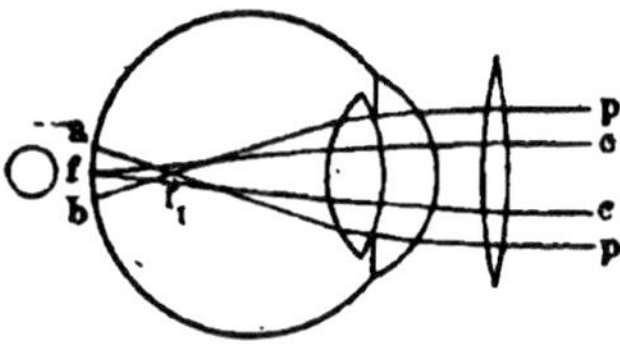

Fig. 4.

By placing the lens before the eye which corrects the hyperopia we increase the refraction, thus eliminating the diffuse central image, but at the same time increasing the peripheral aberration, and therefore also the *area of illumination*, Fig. 4. If the pupil contracted only in *proportion* to the consequent increased light stimulus there would still remain the original diameter of diffusion area. As, however, correction of the refractive error by the lens improves vision, and relieves asthenopia, being *tacit proof* that the aberration is dispelled, it is evident that the pupil must contract *more* than in proportion to the aforesaid light stimulus. Is it not then probable that the pupil contracts *more freely* when accommodation is relaxed?

In controverting this, it would be necessary to refute the following fact pertaining to combined kinetic energies :

When accommodation is in force, the iris is known to be carried for-

* In accommodation, with a standard light placed behind the plane of the eyes, and an approach to them of the paper upon which the test-type is printed, the illumination upon the paper increases in the inverse proportion to the square of the reduced distance between the light source and the test-object. The illumination also varies directly as the cosine of the angle of incidence upon the illuminated surface. (Physical law of Photometry.)

ward,* by pressure from the anterior surface of the lens, which has become more strongly curved. Such lens-pressure, *the iris remaining inactive*, would tend to increase the diameter of the pupil. On this account, greater efforts of the sphincter will be necessary to counteract this action of the lens-surface, when accommodation is present, Fig. 5, than it would with relaxed accommodation, Fig. 6.

For *normal* conditions of innervation the sphincter is known to more than overcome such action on the part of the lens in accommodation, Fig. 5.

Fig. 5.

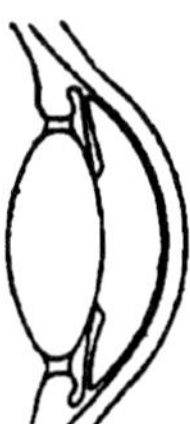

Fig. 6.

If, therefore, our hypothesis is correct, we have found a reason why low degree lenses are of so much benefit in slight hyperopia, and congeneric astigmatism. Furthermore, we are justified in assuming that the sphincter in large pupils does not always adequately respond, *while accommodation is in force*, especially in cases where the optical error is so slight as a quarter dioptry, from the fact that, in the majority of such cases, the patients are young, and often possess amplitudes of accommodation varying between 6 and 14 dioptries.

Patients with such accommodation have so much of it in reserve, even when using the eyes in proximity, that their asthenopia can scarcely be ascribed to an overtaxed ciliary muscle. Are we not then justified in attributing it to possible fatigue of the iris, resulting from its involuntarily prompted, though futile, efforts to exclude peripheral aberration, because of the sphincter's inability, *for some reason*, to contract sufficiently?

It is not recorded that a disproportion of the pupils to the dioptric system of the eyes does ever exist physiologically, but there are many conditions of the nervous system which produce immoderate dilatation of the

*Helmholtz, "Physiologische Optik," page 131, Wien, 1886.

pupils. Such dilatation, *while it lasted*, would tend to oppose the normal association between refraction and the correlated size of the pupil.

In those cases of *normal* pupil, where the perceptive qualities of the retina are good, and the error of refraction is slight, retinal stimulus will prompt contraction of the pupil sufficient to exclude aberration. Is it not probable that, in some cases with *large* pupils, protracted efforts of this kind would result in fatigue of the iris? Might not prolonged ineffectual efforts of the iris to regain equilibrium between its functions, as diaphragm and photostat, account for asthenopia? Or, to put it in another way: Could not that prolonged effort of the sphincter, which would have to be *in excess* of the normal qualitative and quantitative light stimulus, to correct aberration, produce asthenopia?

It need not follow that the iris is incapable of temporarily contracting even to a greater extent than is necessary for the above purpose. This is demonstrated by the extreme contraction of which the pupil is generally capable when exposed to intense light, and the eye is in its static state of refraction.

In hyperopes, we generally ascribe the cause of asthenopia to fatigue of the ciliary muscle, owing to its efforts to exclude the error of refraction by accommodation. The same cannot be said of myopes, whose use of accommodation for such purpose would only render them deplorably more myopic. Their asthenopia can certainly not be ascribed to ciliary fatigue. Some myopes, however, endeavor to improve their vision by compressing the eyelids, which means that they thereby *modify the pupils* to exclude peripheral rays, and the aberration which is heightened by the myopia. In low degrees of myopia and congeneric astigmatism, however, modification of the pupils, by compression of the eyelids, is not sufficiently *delicate* to exclude aberration, *without too great a sacrifice of illumination.* Such patients are therefore more apt to apply for relief from glasses, than those who help themselves by compression of the eyelids, provided this is unaccompanied by asthenopia. In the former cases, we are to suspect that the relief sought is *freedom from peripheral aberration.* The latter also aggravates *photophobia*, which is a symptom frequently complained of in such cases.

The improvement in vision, which the myope, of low degree, with large pupils, secures by the lenticular correction, is practically due to the fact that the peripheral aberration is decreased, through reduced refraction obtained by the concave lens in front, Fig. 7.

The rays emitted from the concave lens, enter the pupil with a divergence counteracting the excessive convergence of the rays which are imper-

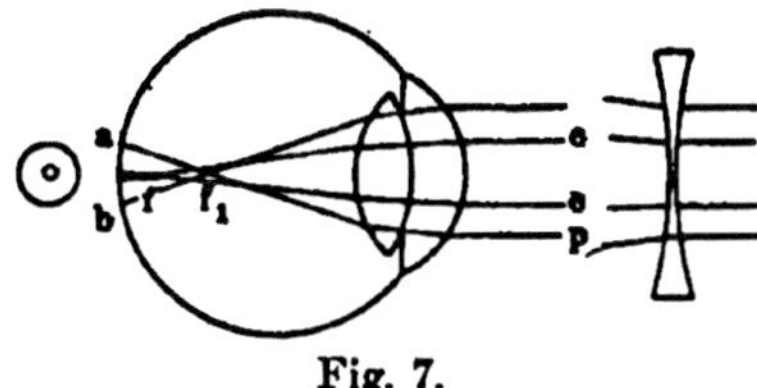

Fig. 7.

fectly focused by the crystalline on the retina behind *f*. The peripheral diffusion area, *ab*, may not, however, always be in such proportion to the central diffusion area as to be fully corrected by the lens which corrects the refractive error in the center. Should it, in the case of a larger pupil, be greater, the patient would merely then select a stronger lens, having its proper effect upon peripheral rays, while its tendency would also be to over-correct the myopia. For a low degree of myopia, this would scarcely be appreciable, since very little difference in refraction is experienced in the actual centers between lenses of a quarter and a half dioptry.

In those cases where the quarter-dioptry lens seems to relieve asthenopia it will generally be found that the pupils are comparatively large. This is especially noteworthy in those cases where simple myopes of low degree are benefited by wearing their weak distance corrections for *reading*, and which can serve no other *needful* purpose than to eliminate peripheral aberration.

So far, we have no means of ascertaining the size, or that variation of the pupil which is necessary to establish the proper harmony between refraction, accommodation, illumination, and freedom from aberration. The intuitive discrimination, which accompanies experience, is at present our only guide.

In refractive errors of low degree, which are relieved by lenticular correction, the retinal perception is usually also very keen, thus increasing stimulus to contraction of the sphincter, while the correction in such cases frequently improves vision to $\frac{6}{3}$, which is far above normal.

The larger the pupil, the more pronounced will be the improvement in visual acuteness obtained by low-degree corrections. The quarter-dioptry lens rarely proves of benefit when the pupils are small.

Again, patients frequently wear such glasses for a time, relieving their asthenopia, and ultimately lay them aside, without feeling the necessity of their further use. Examination will nevertheless reveal the fact that *the optical error has not changed.* Why then should asthenopia exist at one time, and not at another, for an *invariable* hypermetropic astigmatism for instance, if the fatigue in the first instance had only been due to that of the ciliary muscle?

Closer examination, however, will frequently show that the pupils appear to be smaller at the time the patient has discarded his glasses than when they were prescribed. The pupil being the only member seeming to have undergone a change, are we not justified in suspecting the iris, by reason of disturbed innervation, as having been at least implicated in the cause of asthenopia?

THE TYPOSCOPE.

Revised reprint from The Keystone, 1897.

It is commonly understood that visual acuteness depends upon the perceptive functions of the retina, as well as upon the size of the image projected upon it, and which is limited by the visual angle subtended by the object at the nodal point within the eye. To the exclusion of all other considerations, the ability to discern objects therefore primarily depends, first, upon contrast in light intensity, involving also the color sense; and second, upon the size of the object viewed. Therefore, the greater the contrast between the color of the object and the background, the more readily will an object of any given size be distinguished. Thus it is frequently observed that a visual acuteness of $\frac{6}{9}$, with diminished illumination, is raised to $\frac{6}{6}$ with a maximum illumination, as a result of heightened contrast between the type and its background. In cases of ametropia and amblyopia it is, however, also frequent that increased illumination reduces the definition, owing to a superabundance of extraneous light, which serves to reduce the contrast within the field of actual fixation. In optical instruments it is found practicable to exclude extraneous light by means of a diaphragm of suitable aperture, and it is even possible to increase the definition, through limiting the field in an inferior instrument by further reducing the size of this aperture. Thus it is that the pin-hole disk heightens the visual acuteness in ametropes who view objects at a distance through it. While the same proportionate improvement can be obtained in a similar manner at finite distance, yet it would be exceedingly difficult to accurately place the pin-holes before the pupils of both eyes for reading binocularly. To obviate this impracticability, while still securing an unimpaired field of fixation, the typoscope, as here described, seems in many instances to effectively serve its purpose.* It consists of a rectangular plate of hard rubber, or black cardboard, 7 by 2¼ inches, provided with an

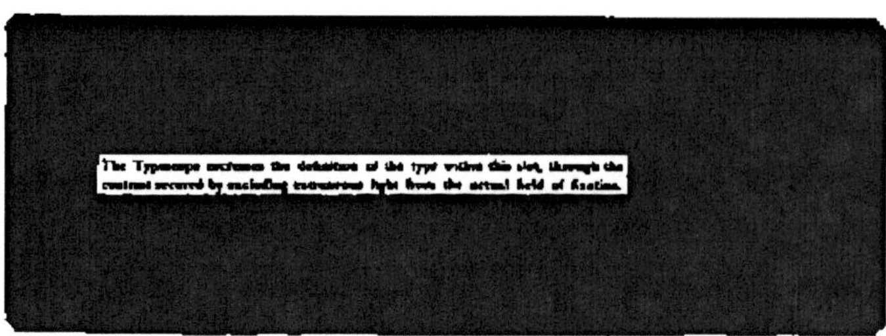

One-half of Actual Size.

aperture 4¼ by ⅜ inches, centrally located, though laterally displaced so as to leave sufficient of the plate, two inches, to be conveniently held between the thumb and fingers, when it is placed upon the book or paper, and while it is being slid down over the column in reading. The central aperture is just deep enough to allow two lines of brevier type to be viewed at a time, and wide enough to take in the width of an average column of type, as shown in the diagram. The author has found it to be especially serviceable to cataract patients and amblyopes wearing high corrections. The former, who notably suffer greater impairment of vision from extraneous light, are invariably enabled with their glasses to read the smallest type by the aid of the typoscope, which excludes all light reflected from the surface of the paper, except that which actually affords them the necessary contrast between it and the type within the slot. The device is exceedingly simple, inexpensive, and easily carried in the pocket. Its utility is easily demonstrated by first ascertaining the size of the smallest type which the patient reads with glasses, and then allowing the patient to use the typoscope in addition to them, for the purpose of ascertaining whether smaller type can be read, or not. Even in the latter case it has been the author's experience that patients using the typoscope claim to read with less sense of confusion.

* "I am delighted with the typoscope. It rests on sound physiological principles, and will benefit many people."—H. Knapp, M. D., New York.

THE CORRECTION OF DEPLETED DYNAMIC REFRACTION.

(PRESBYOPIA.)

Revised reprint from the "Optical Journal," June, 1895.

To render this subject fully comprehensive, we shall first briefly describe Donders' Chart of the Amplitudes of Accommodation in Emmetropia, and in which the "near points" are represented by a continuous curved line drawn diagonally through its field. The successive ages, between 10 and 80 years, are therein pointed off upon the uppermost horizontal line, which also represents the plane of the eyes. On the right hand margin, the distances of the near points from the eye are placed opposite to the horizontal lines which intersect the verticals apportioned to the various ages. On the left hand margin the same horizontals are numbered in dioptries of refraction, counted from the zero-line (o) below, which is supposed to be at infinity. The latter therefore corresponds to the refraction of the eye when at rest—its *static* refraction. Between 50 and 80 years of age it will be noted that the line representing the static refraction curves slightly downward at the right, showing that the punctum remotum becomes negative, that is to say, the refraction of the emmetropic eye *acquires* hyperopia from age. The changes in the refraction of the eye which are accomplished by efforts of the ciliary muscle, are termed its *dynamic* refraction or power of accommodation. The amplitude of accommodation, at any age, is equal to the number of ruled spaces between the zero-line and the curved diagonal line above, that is to say, equal to the difference between the static and dynamic refraction. Therefore, if we represent the range of accommodation by a, the static refraction by r, and the dynamic refraction by p, we have

$$a = p - r$$

as the formula for the range of accommodation.

In emmetropia, the eye is adapted to infinity, so that the static refraction is *nil*, and therefore

$$a = p$$

which means that the accommodative effort at the punctum proximum is equal to the amplitude of accommodation.

From early childhood the accommodation is shown to gradually decrease, though it only becomes manifest to persons as they approach the age of 45, when the punctum proximum about reaches the accustomed distance for reading. The patient then discovers a loss of distinctness in reading, and experiences a desire to hold his book at a slightly greater distance than is desirable. This serves to show that so long as there is sufficient power of accommodation in reserve, that is to say, capacity to adapt the eye inside of the limit of the usual reading distance, there will be no asthenopic symptoms from what is commonly called Presbyopia, if by the latter we only mean to designate a recession of the near point beyond the accustomed finite occupation distance. Experience has also shown that the reserve accommodation should exist in a more or less definite proportion to the amount required for a given finite distance. It would prove futile, for instance, to prescribe glasses at the age of 45, which would artificially afford the patient as much refraction as he possessed dynamically at the age of 35.

Landolt* says that "the accommodative effort is not to be measured by any fixed standard, but finds its expression in *the relation between the effort produced and the entire amount of accommodative power at disposal.* It follows from this that the reserve fund of accommodation must have a *relative*, and not an absolute, value; it must be a quota of the range of accommodation." He has found, experimentally, that "a continued effort of the ciliary muscle is practicable only when it calls for but two-thirds, or at the utmost three-fourths of the total power of accommodation."

*The Refraction and Accommodation of the Eye, page 339, by E. Landolt, M.D., Paris; translated by C. M. Culver, M.A., M.D., Philadelphia, 1886.

In the accompanying chart, the author has therefore interposed the corrections for depleted accommodation, by assuming that three-fourths of the

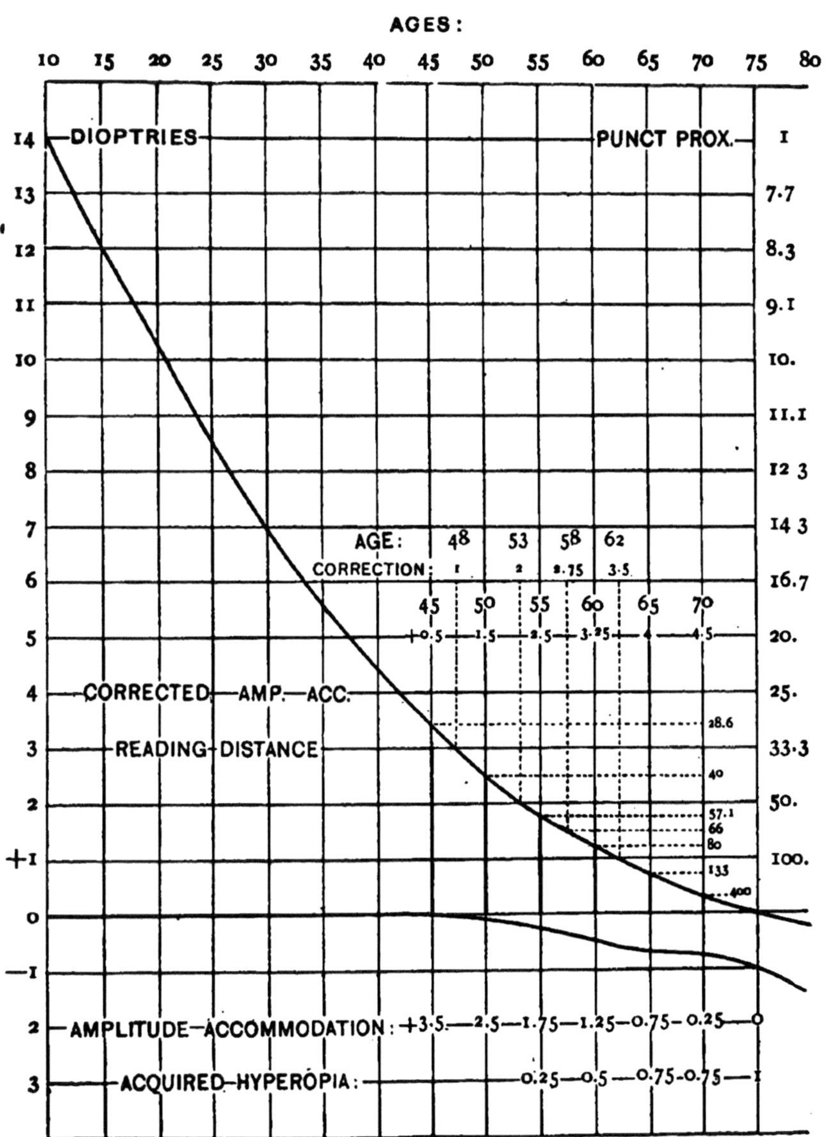

total accommodation required at 25 cm. is used in reading at 33⅓ cm. distance, because the corrections then more closely correspond to Donders' table, and also for the reason that the author has found it more satisfactory to thus calculate in practice. For instance, the emmetrope, whose amplitude of accommodation is 2.5 *D.*, should have 3 *D.* of refraction to read at 33⅓ cm., but, as he must also have one-third of the accommodation required at this distance in reserve, he should have 4 *D.* of refraction to read comfortably. Since his dynamic refraction is 2.5 *D.* we therefore give him + 1.5 *D.* glasses to supply the deficiency. If his range of accommodation is normal, we find, by the chart, that he should be 50 years of age. In this manner the author frequently estimates the ages of persons with surprising accuracy.

The lenticular corrections, inserted in the field of the chart, between the ages of 45 and 80 years, added to the corresponding amplitude of accommodation, *less the acquired hyperopia* noted beneath, are found in each instance to amount to 4 *D.* Consequently, in any case, the requisite 3 *D.* to read at 33⅓ cm. represents three-fourths of the total power supplied both dynamically and artificially, by lenses.

Of course, the corrections in the table, are only applicable in such cases where the reading distance is no greater than 33⅓ cm., and wherein the amplitude of accommodation is found to correspond to Donders' determinations.

The chief value of the chart therefore exists only in the fact that it serve, to show, by comparison with any given case under examination, to what extent its amplitude of accommodation differs from the accepted normal state as determined by Donders. This cannot be too highly estimated, however, for Landolt* says: "Donders' diagram corresponds so perfectly to the natural condition of things, that, in every case where the amplitude of accommodation is less than is indicated thereon, we may safely diagnose a weakness of accommodation, and, in case of any considerable difference, we may admit a paresis of this function." A knowledge of the patient's condition of health and habits will of course assist greatly in arriving at a definite decision.

The *punctum proximum* is that point located at finite distance, at which

* Landolt's work, page 556.

the patient is still able to distinctly see small printed characters, such as diamond type, dots, or fine lines. The size of the test-objects should, of course, be in proportion to the visual acuteness of the eyes, since there are persons, who, though possessing a good range of accommodation, cannot read fine print at any distance, simply because their visual acuteness is insufficient. Therefore, in cases where the visual acuteness is $\frac{6}{9}$, or less, it is preferable to use more heavily executed test-objects. For a normal visual acuteness, the author finds it convenient to use the characters here shown and which are engraved on a circular disk mounted in a handle, to be held by the patient. On the obverse side of the disk, the same, though more heavily drawn figure, *a*, is used when the visual acuteness is subnormal.

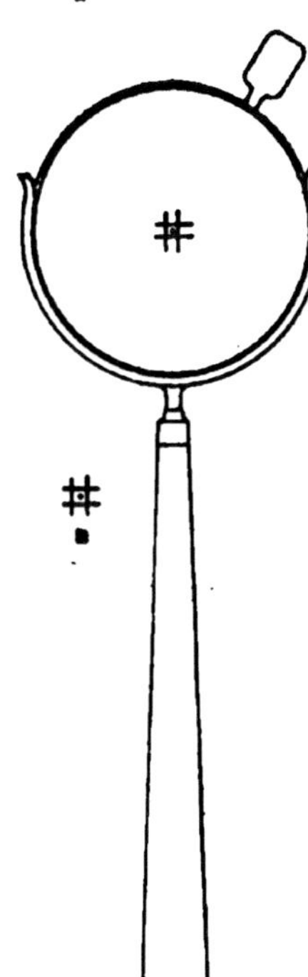

To determine the position of the punctum proximum, the patient is requested to binocularly fix the central dot of the test-object at about his usual reading distance, and to gradually draw it nearer to the eyes until it just begins to blur; the latter effect being made more noticeable to the patient through the tendency of the dot to fuse with the lines of the square which encloses it. The punctum proximum is reached the instant before the blurring is observed. This experiment should be repeated several times, and it is generally also advisable to verify it by moving the dot slightly nearer than the punctum proximum, having the patient fix the dot attentively while it is gradually withdrawn to the position where it again appears sharply defined. By using a tape, which is graduated to dioptries of refraction, we may read directly from its graduations, the amplitudes of accommodation in emmetropia, provided the distance is measured from the cornea to the test-object, on the median line. The same procedure will apply in any case of *ametropia*, when the distance glasses which correct it are worn at the time the above measurement is made. In the latter case, calculation will of course be greatly simplified.

To those familiar with the art of fitting glasses it is, in most instances, comparatively easy to determine the proper distance glasses, as well as for them to predict, with reasonable certainty, that the lenses will be worn with comfort. But when the case is complicated with a loss of dynamic refraction, and unless extreme caution is used, there is great danger in giving the patient an over-correction, thus ultimately making a change in the prescribed reading glasses necessary.

It is therefore of great importance to the optical practitioner, particularly if the cost of a subsequent change in the reading glasses is to be borne by him, that he should be able to predict the correctness of the reading glasses with the same degree of certainty that he feels in respect to the glasses which he prescribes for distance. In prescribing for depleted accommodation there are at present two methods in vogue:

1. The *impirical* method; that of prescribing the glasses which have been accorded to a given age by Donders in his table.

2. The *physical* method; that of locating the punctum proximum in each individual case, and using this as the basis for calculating the loss of accommodation which is to be compensated for.

The latter is the only accurate and reliable method, yet even in applying it we are frequently hampered by the patient's indecision as to the distance at which he habitually performs his near work. In the practitioner's office the patient may indicate the distance as being thirteen inches, whereas at his own occupation he perhaps finds that it is twenty. To avoid this feature of uncertainty as much as possible, it is consequently prudent to have the patient state the nature of his occupation, and to have him assume his accustomed position of the head, arms and body when engaged in near work. Then measure the distance from the eye, during convergence to the median line, by the dioptral tape, and note it as the *desired* reading distance, which is after all the real and only distance to be considered in the calculation for reading glasses.

As an example, let us take an emmetrope who has a range of accommodation of 2 *D.*, and who indicates, by the aforesaid measurement, that he desires to see at the distance which corresponds to a refraction of 2.25 *D.* by the tape. As this amount should represent only three-fourths of the total refraction, to allow him one-fourth in reserve, it follows that the total refraction should be 3 *D*. As he is capable of contributing 2 *D.*, dynamically, we give him + 1 *D.* glasses to supply the deficiency for reading.

Therefore, to ascertain the reading glasses for any given case of *emmetropia*, we have only to follow the simple rule:

Increase the refraction corresponding to the desired reading distance by one-third, and subtract therefrom the patient's amplitude of accommodation.

It happens occasionally, as in the above example, that the punctum proximum is too far distant to enable the patient to see the test-object *distinctly*. In such cases it is convenient to assist the patient by a lens which will enable him to do so. Let us suppose that we have assisted the patient impirically by a 1 *D.* lens, when he will, by the use of his 2 *D.* of accommodation, be able to see the test-object distinctly at the 3 *D.* distance. By deducting the 1 *D.* lens we then find his amplitude of accommodation, which is 2 *D.*, and proceed by the rule given.

In *ametropia*,* provided it is corrected by the distance glasses, which then virtually render the patient emmetropic in accommodation, we proceed by the same rule. The full reading correction will in this case be equal to the amount found by the rule, *plus* the distance correction. Should the accommodation be insufficient to definitely locate the punctum proximum, the patient, here too, is to be assisted by a lens which will likewise have to be deducted to ascertain the amplitude of accommodation. An exception to an application of the rule is found in those cases of myopia where the punctum proximum is nearer than the *desired* reading distance. In such cases it is customary to deduct the refraction corresponding to the desired reading distance from the distance correction. In some instances, where the amplitude of accommodation is considerable,

---

* When the ametropia is not corrected, that is to say, when the distance glasses are not worn during the measurement of the range of accommodation, we must resort to the formula: $a = p - r$. In hyperopia the punctum remotum is behind the eye, therefore the refraction is negative, so that

$$a = p - (-r) = p + r$$

which means that the range of accommodation is equal to the refraction at the punctum proximum, plus the refraction of the lens which corrects the hyperopia.

In myopia the amplitude of accommodation is equal to the difference between the refraction at the near point, and the refraction of the lens which corrects the myopia. For a more exhaustive discussion of this subject, the reader is referred to Dr. Landolt's work, in which the physical portion is treated at greater length and more lucidly than in any other medical publication which has come to the author's notice.

and more especially in young persons, the patient will prefer to use his distance correction for all purposes.

From this discussion it must be evident to anyone proficient in the practice of optometry that, as a matter of fact, greater skill and knowledge is required to scientifically determine the proper glasses for reading, commonly known as presbyopic corrections, than is needed to ascertain the lenticular requirements for distance. Nevertheless, some oculists have expressed the opinion that opticians should only be permitted to adapt glasses for presbyopes. If the same medical gentlemen were better informed in optics they would undoubtedly deny opticians all rights in the matter.

CPSIA information can be obtained at www.ICGtesting.com
Printed in the USA
LVOW09s1325090316

478450LV00019B/235/P